T0173194

# THE EVERYTHING BLUEPRINT

# THE
# EVERYTHING
# BLUEPRINT

## THE MICROCHIP DESIGN
## THAT CHANGED
## THE WORLD

### JAMES ASHTON

HODDER &
STOUGHTON

First published in Great Britain in 2023 by Yellow Kite
An imprint of Hodder & Stoughton
An Hachette UK company

2

A CIP catalogue record for this title is available from the British Library

Hardback ISBN 9781529394054
Trade Paperback ISBN 9781529394061

Typeset in Fairfield LT Std by Manipal Technologies Limited

Printed and bound in Great Britain by Clays Ltd, Elcograf S.p.A.

Hodder & Stoughton policy is to use papers that are natural, renewable and recyclable products and made from wood grown in sustainable forests. The logging and manufacturing processes are expected to conform to the environmental regulations of the country of origin.

Hodder & Stoughton Ltd
Carmelite House
50 Victoria Embankment
London EC4Y 0DZ

www.hodder.co.uk

*To Mum and Dad, for the best start*

# CONTENTS

# PREFACE

I was unashamedly greedy in writing this book. In search of the biggest business story I could find that was anchored, like me, in the UK, but touching every part of the globe, I quickly alighted on microchips, even before the semiconductor shortages of 2021 propelled these tiny devices to the top of the news agenda.

It is a story that has everything: brilliant science, dogged entrepreneurialism, great rivalries, huge financial risk and reward, and technology that delights consumers and spooks politicians.

And the perfect way to tell it all was through Arm, a young company by British standards whose microchip designs operating at low power and selling at low cost have populated billions of machines and helped to break new ground in dozens of industries.

During the research process I recalled an encounter with Google co-founder Larry Page, back when I was writing for the *Sunday Times*. He likened his business, which was then largely an internet search engine, to making toothbrushes. As he hunted for the next innovation, Page aimed to create something simple that people would use every day from which he could take a small cut, just as he had monetised mouse clicks with advertising so effectively.

The ubiquity that means Google still dominates its market by carrying out billions of searches every day has a direct read-across to Arm, which has also created something too convenient, too reliable and too cheap for users to hunt far for an alternative.

When the UK craves more home-grown technology leaders, it is instructive to discover how one company manages to serve – and serve simultaneously – Apple, Amazon, Samsung, Qualcomm, Alphabet, Huawei, Alibaba, Meta and Tesla – modern industrial titans and very often arch competitors.

And, in an increasingly politicised industry, where governments are spending billions of dollars to capture a slice of the strategically important microchip supply chain, it feels important to understand how Arm came about with little more help than some profits left over from an early-1980s state-sponsored effort to boost computer literacy. Arm takes its place in an industry governed by paradox, where great litigation runs in parallel with great collaboration. However logical and methodical the processes of computing, success depends on near religious fervour for one technology over another, the magic of marketing, momentum with customers – and a dash of luck.

This book is a study of corporate Darwinism over decades: how IBM grabbed hold of the personal computer market in the 1980s but the real winner was Intel; how Nokia beat Ericsson and Motorola to popularise and dominate mobile phones in the 1990s; how Apple, Samsung and Huawei overthrew them all in the 2000s; how Amazon in the 2010s captured cloud computing; and how Arm pulled ahead of numerous other microchip architectures to prosper magnificently. At almost every stage, the answer to 'why?' lies with the tiny components behind the brands, buried in the devices, doing the hard work.

I have presented the Arm story in three parts. First, the origins of its core technology, the company that commercialised it and the vital role played in the mobile-phone revolution it could have been made for. Second, how Arm branched out, leaping from 'dumb' phone to smartphone, and the success that attracted stiff competition from US chip giant Intel and an offer it could not refuse from SoftBank, the Japanese investment firm. Third, forging new markets in low-energy sensors and high-end data centres, and navigating the pitfalls its pervasiveness has brought.

Laced throughout is the long-standing relationship with Apple, kicked off during co-founder Steve Jobs' 11-year absence from the firm, which contributed to Arm's formation, the breadth of its commercial success and appeal to its acquirer SoftBank, as well as Apple's return to financial stability and its transformation under Jobs from a computer company into a consumer electronics leader.

Such is the range of the microchip industry, some chapters step away from the Arm story to add context: how chips originated in the United States but production quickly gravitated to Asia; why China has sought to become self-sufficient in chips and the US has sought to stop it; how numerous nations are now jostling to produce their own.

Two more companies deserve walk-on parts for their indispensability to this ecosystem that helps to explain Arm's own. The most intricate chips in the world are manufactured by Taiwan Semiconductor Manufacturing Company (TSMC), without whose supply both the US and China would struggle. Those intricacies are etched onto silicon using machines built by ASML, a Dutch company that has fought off American and Japanese rivals over decades to become the undisputed market leader. Both also find themselves in the crosshairs of the bitter battle for semiconductor supremacy that is being fought between the US and China.

In writing this story, at times I have felt like a chipmaker, poring over intricate plans and etching ever more information into a relatively minute space. This is a mainstream telling of a complex history and geography, honouring the science while trying not to be blinded by it. It is just technical enough, I hope, without slowing the narrative, and careful not to narrow from the awesomeness of what semiconductors have made real.

More than anything, beyond the torrent of electrons and some magical machinery, I discovered it was a story of people. We live in an era with a rising tide of artificial intelligence – for example, ChatGPT – that makes it possible to sense the point at which computers will take over. Nevertheless, innovation remains a human pursuit, with trust, respect and often friendship at its heart. The ideas behind the next breakthrough will be just as fluid as the last: shared, stolen, enhanced by colleagues and competitors, no matter who employs them – or where.

The people who won't be told what is impossible continue to power remarkable progress that the whole world stands to benefit from. At Arm and beyond, they must not be constrained from continuing.

James Ashton
January 2023

# Chapter 1

# EVERYTHING, EVERYWHERE: HOW THE MICROCHIP TOOK OVER

## A Silicon Shield

A little after 10.45pm local time on 2 August 2022, a Boeing C-40C jet touched down on the tarmac at Songshan airport in Taipei, the capital of Taiwan.

Since leaving Kuala Lumpur, Malaysia, seven hours earlier, the US Air Force craft had taken a circuitous route. But as it headed towards the Indonesian part of Borneo and then turned north to skirt the eastern edge of the Philippines, all eyes were on it. FlightRadar24, a flight-tracking website, registered 2.9 million viewers during its journey, making it one of the most closely tracked flights of all time. And most of those that logged on knew precisely why the pilot had chosen to avoid the most logical, direct route over the South China Sea.

Emerging into the night, an elderly woman dressed in a smart pink trouser suit and face mask grasped both handrails as she gingerly descended into the melee on the runway. There, Nancy Pelosi, the Speaker of the US House of Representatives and her country's second-most powerful politician, was greeted by Taiwan's foreign minister, Joseph Wu, and the US representative in Taiwan, Sandra Oudkirk.

It was an arrival freighted with meaning. In anticipation, Hong Kong's Hang Seng index had fallen 2.5 per cent, China's Shanghai Composite dropped by 2.3 per cent and yields on 10-year US Treasury notes touched a four-month low.

For 25 years, there hadn't been a US visit of this magnitude to Taiwan, an island that China regarded as its own. Under its 'One China' policy, Washington had been careful to acknowledge Beijing's position that there was only one Chinese government, and it did not have formal diplomatic relations with Taiwan, although its unofficial contact and 'strategic ambiguity' included supplying weapons and offering tacit support in case of emergency.

But the status quo was fraying. The Chinese President, Xi Jinping, had stated that 'reunification' with Taiwan 'must be fulfilled' and warned US President Joe Biden not to overstep the mark, informing him on a recent call that 'those who play with fire will perish by it'. In the days running up to Pelosi's rumoured arrival, China massed warships and planes near the median line, an unofficial border in the waters between China and Taiwan, and tanks were mobilised on the mainland.

Undeterred, on her whistle-stop tour Pelosi declared that US determination 'to preserve democracy here in Taiwan and around the world remains ironclad'.[1]

There wasn't just political tension. The situation was overlaid by vast economic risk too. Over many years, the build-up of China's military might in the region was matched only by Taiwan's growing pre-eminence in producing the most valuable commodity of the 21st century: microchips.

Geopolitical ructions of the past stemmed from the pursuit of resource-rich land or religious ideology. But in the digital age, these tiny slivers of silicon that powered cutting-edge weaponry, as well as smartphones, cars and medical equipment that had taken on new importance since the devastating Covid-19 pandemic, were the ultimate prize.

In less than two generations, an island with a land mass slightly larger than Maryland and less than twice that of Wales had grown to account for 92 per cent of the world's most advanced chips. They are defined as those produced using manufacturing processes carried out at 10 nanometres (10nm) or less.[2] That so-called 'node' is a measure of the spacing between transistors packed on the chip, or roughly four times the width of a strand of human DNA. The rest originated from South Korea. Neither the US, the inventor of this industry decades earlier, nor China, which was striving for self-sufficiency from the US, could cope without Taiwanese output that fulfilled the goods their factories made and consumers and businesses faithfully bought.

Demonstrating how closely entwined the two opposing sides were, China consumed 60 per cent of global chip output, but about half of that volume departed the country in finished goods, and many of those were destined for the US, the product of an intricate global supply chain that had been established in better times.

All of this explained why Pelosi, alongside the Taiwanese President, Tsai Ing-wen, found time to sit down for lunch with Mark Liu, the chairman of Taiwan Semiconductor Manufacturing Company (TSMC), and its talismanic founder, Morris Chang.

———————

TSMC was the world's largest contract chipmaker, making chips to the specification of its customers but not designing its own. In 2021 it manufactured 12,302 different products using 291 distinct technologies for 535 different buyers. Its largest customer was widely thought to be Apple, which accounted for roughly one-quarter of TSMC's output for its iPhones, iPads and watches, but it also served a who's who of the industry – Qualcomm, Nvidia, NXP, Advanced Micro Devices (AMD), Intel – all of which supplied chips to numerous market ends. TSMC also made chips for US military equipment, including, reportedly, F-35 fighter jets and Javelin missiles.

In the past, TSMC's strategic importance was regarded as a 'silicon shield' sufficient to protect Taiwan from Chinese attack and maintain US backing. Now military experts were not so sure. A war between the US and China would be devastating, and potentially redraw the world order. But shutting down Taiwan's chip factories would cripple the world economy. In this tinderbox, both outcomes were possible.

Most of TSMC's fabrication plants – known as fabs – were directly in the line of fire. Clustered at Hsinchu Science Park, along Taiwan's west coast facing China, they were close to the 'red beaches', so-called because they were likely landing sites for Chinese forces.

But if China invaded, there was no saying it could secure the fabs intact. And even if it did, running the production lines was not as simple as flicking a switch. They needed Taiwanese expertise for that, plus thousands of support engineers based at partner companies all over the world, who communicated with Taiwan-based staff remotely, often via augmented reality. TSMC's Liu warned that an invasion would bring about the 'destruction of the world's rules-based order' and render its factories as 'non-operable'.[3] And, in the unlikely case the Chinese could overcome that hurdle and keep the facilities running, their output would quickly be cut off from the rest of the global supply chain.

'How do we get it resolved?' Liu said. 'It's very simple. Keep Taiwan safe.' He added: 'If you don't do that you have to spend hundreds of billions of dollars, (over) maybe 10 or 15 years, before you get back to this point.'

A year before the Pelosi trip, veteran semiconductor industry analyst Malcolm Penn sketched out the catastrophic impact of an attack. 'Chip inventories would quickly become exhausted and end equipment production lines everywhere would grind to a halt within a matter of weeks, even days,' he wrote in his FutureHorizons briefing. 'The near instant impact on global trade and the world economy would be orders of magnitude greater than the 2008 Lehman Brothers crash or the 2020 Covid-19 lockdown.'[4]

## A 21st-Century Horseshoe Nail

Early morning on Thanksgiving in November 2021 and the line outside GameStop in Murphy Canyon, San Diego, snaked into the car park. Many retailers had chosen to close for the holidays but this store had good reason to open its doors.

At the front of the queue, Joland Harper, in green beanie hat, black T-shirt and sporting a nose ring, had been waiting since 6am. 'I've got my snacks here, got the umbrella, got the comfy chair – so I was ready for the wait,' he told the local CBS news channel with a smirk.[5]

Harper and the small crowd stood behind him had turned up on this bright and cool day for just one reason. Word had got out that GameStop had taken delivery of 24 Sony PlayStation 5 consoles, the must-have gadget that had been resolutely hard to get hold of ever since its release the previous Thanksgiving.

Consumers used to the instant gratification of online ordering and delivery had been going to great lengths to track down the machine. Many who rarely bothered with bricks-and-mortar shops resorted to forming an orderly queue for a change. The

Twitter account @PS5StockAlerts, dedicated to tracking deliveries, gained a million followers. And in one normally sedate Tokyo department store, shoppers rushing to grab the machines resulted in the police being called to restore order.

The reason for all this inconvenience was hard to compute. A global shortage of microchips, the low-profile but essential power inside games consoles – as well as every other device imaginable – was wreaking havoc on production lines and with warehouse supplies around the world. In many instances, the absence of a single critical chip that cost less than a dollar prevented the sale of a device that cost many thousands of dollars.

Demand had grown steadily over the past decade as more chips were packed into each product. But it skyrocketed when consumers' lockdown lifestyle called for new electronics for work and play. Supply was stuck when chip factories were forced to down tools during the pandemic, especially in the 'back end' stages of production such as packaging and testing, which remained fairly labour-intensive.

The problem was widespread. Between the start of January and October 2021, microchip lead times – the gap between when a semiconductor was ordered and when it was delivered – ballooned from 14 weeks to 22 weeks, according to Susquehanna Financial Group. Waits for some specialist components were longer.[6]

The snafu served as a reminder that chip supplies were strategically important, the industry relied heavily on a handful of major players, there wasn't much slack in the system – and the cost of entry for anyone else was vast. From flying under the radar for so long, chipmakers, designers and essential tool providers were catapulted into the public eye in the same way the 1970s oil crisis focused attention on that industry's production and practices. There were several reasons behind Pelosi's trip to Taiwan, but one of them was clearly to ensure US consumers like Harper would never have to queue in line again.

In July 2021, Sony was able to celebrate that the PS5 had passed the 10 million sales milestone a few weeks faster than its predecessor, the PS4, but the Japanese electronics giant would have sold many more if it could have kept up with demand.

'While we continue to face unique challenges throughout the world that affect our industry and many others, improving inventory levels remains a top priority,' said Jim Ryan, the chief executive of Sony Interactive Entertainment.[7] By March 2022, Sony had shifted some 19.3 million consoles, some 3.3 million fewer than had been hoped at that stage.

———————

Meanwhile the car industry, which scaled back chip orders when it thought vehicle sales would slow, scrambled to catch up with demand and felt the squeeze keenly when their capacity was taken up by other industries. With up to 3,000 chips per model – powering everything from braking systems to in-car entertainment – the world's biggest auto makers were forced to operate fewer shifts and extend holiday closures. What chips they obtained were allocated to higher-profit vehicles, but that was little comfort to Volkswagen's board chairman Herbert Diess, who said his group, whose brands included Audi, Lamborghini, Seat and Skoda, was in 'crisis mode'.[8]

In September 2021, the consulting firm AlixPartners forecast shortages would cost the global industry $210bn in lost revenues that year and 7.7 million fewer cars would be produced, up from its May prediction of $110bn and 3.9 million fewer units.[9] Second-hand car prices soared.

The industry's dependence on chips was only going to increase. Electronics would account for an estimated 45 per cent of a car's manufacturing cost by 2030, according to a Deloitte report, up from 18 per cent in 2000. Over the same period the cost of the semiconductor-based components used in those electronics was estimated to quadruple to $600.[10]

The shortage had companies, industries and countries all jostling for position, hoarding what they had and ordering more than they needed. Players in niche but vital industries took to pleading in public for their supply. 'Due to the urgent need for these in the medical technology industry, representing just 1% of the total supply, we call for chip allocations to be prioritized to a level that enables the industry to meet the medical-device manufacturing demands of today,' wrote Frans van Houten, the chief executive of Dutch electronics firm Philips, in June 2022.[11]

The crisis turned thoughts from chip usage today to security of supply tomorrow. Rather than merely carrying out tasks as requested, the most advanced chips powered artificial intelligence (AI) that would churn through streams of data and make decisions faster than any human.

In his 2005 book *The Singularity Is Near*, the futurist Ray Kurzweil predicted that singularity – the point at which a computer's abilities would overtake those of the human brain – would occur in about 2045. Experts have since predicted it could be much sooner, given indicators such as the defeat of Lee Sedol, the world champion in the ancient Chinese board game Go, by the AlphaGo computer program in March 2016.

Whatever the date, AI promised to revolutionise every aspect of life – including warfare – which made nations nervous. In a report submitted in March 2021 to US Congress, the bipartisan

National Security Commission on Artificial Intelligence warned starkly: 'America is not prepared to defend or compete in the AI era.'

The report tried to capture the game-changing nature of AI, beyond any single technological breakthrough. To encapsulate an AI-powered future, it leaned on words by the great inventor Thomas Edison as he tried to describe the potential of electricity. 'It is a field of fields,' he said. 'It holds the secrets which will reorganise the life of the world.'

The US was exposed because it no longer manufactured the world's most sophisticated chips. 'We do not want to overstate the precariousness of our position,' the report said, 'but given that the vast majority of cutting-edge chips are produced at a single plant separated by just 110 miles of water from our principal strategic competitor, we must re-evaluate the meaning of supply chain resilience and security.'[12]

One month after his inauguration, on 24 February 2021, President Biden labelled the shortage of computer chips as 'a 21st-century horseshoe nail', recalling the proverb that suggested small acts or items could have vast, unforeseen consequences. 'For want of a nail the shoe was lost,' it begins, concluding with the lines: 'For want of a battle the kingdom was lost, and all for the want of a horseshoe nail.'

The message was clear: tiny microchips were the bedrock on which the modern knowledge economy was built. Two years of shortages translated into $500bn of lost revenues across numerous industries that relied on chips, one analysis found.[13] Chips translated into earning power and, increasingly, political power. The US, cradle of microchip invention, still accounted for about half of global industry revenues, but its share of manufacturing capacity, which had been 37 per cent in 1990, had slumped to just 12 per cent.[14]

'We need to stop playing catch-up after the supply chain crisis hit,' Biden said, holding up a chip smaller than a postage stamp. 'We need to prevent the supply chain crisis from hitting in the first place.' He ordered a 100-day review of four vital products, including semiconductors, and a longer-term review designed to 'fortify our supply chains at every step'.[15]

Thirty-two years earlier, the same technology was on the mind of one of Biden's predecessors, Ronald Reagan. His first overseas speech after leaving office was delivered in the Gothic splendour of London's Guildhall in June 1989. It was soon after China had crushed the student-led demonstrations held in Beijing's Tiananmen Square. In November that year, the Berlin Wall would fall.

'Information is the oxygen of the modern age. It seeps through the walls topped by barbed wire, it wafts across the electrified borders,' Reagan said. 'The Goliath of totalitarianism will be brought down by the David of the microchip.'[16]

Sadly, it hadn't panned out like that. In fact, a new Cold War was brewing – with the supply of microchips at its heart.

## Industrial Rice

The printing press educated, the light bulb illuminated, the agricultural plough changed diets and landscapes, the motor car broadened horizons. But the microchip could be the most remarkable invention of them all.

It stakes a claim among centuries-old breakthroughs – making fire, handwriting, the wheel, the compass – that have transformed how humans live and learn. A tiny device that makes 'dumb' items 'smart' today extends its influence over the fabric of society in which individuals cannot complete simple tasks without at least basic computational support.

Chips are quietly at work everywhere. Even in a year supposedly riven with shortages, a record 1.15 trillion of them were built and sold in 2021, installed in computers, phones, TVs, cars, refrigerators, oil pipelines, security systems, data centres, pacemakers, pets, toys, toothbrushes, nuclear missiles and more.[17] That is a cache of 125 tiny new devices for every human on the planet, on top of the trillions already out there.

Since the integrated circuit was invented on a sliver of semiconducting silicon in the late 1950s, these mini-machines have become omnipresent in modern life. Microchips – also known as semiconductors, integrated circuits, systems-on-a-chip, microprocessors and microcontrollers – power connectivity and creativity, enabling human endeavour to go further, faster, better.

Opportunities have expanded as components and prices shrank. The industry has tried relentlessly to do more for less. It is the 'global race' writ large – itself invoked by politicians the world over as they urge their populace to work faster and smarter and more efficiently or else face obsolescence at the hands of nimbler economies. Microchips are a vital staple that feed progress. No wonder Yun Jong-yong, the one-time vice chairman of Samsung Electronics, one of the world's largest chip producers, labelled them as 'industrial rice'.

They have borne the internet to ubiquity, generating untold wealth – and not just for the tech billionaires whose social media software skates over chip-laden hardware. Over decades microchips have catalysed a sustained increase in productivity. Today, they are a $550bn-a-year industry that drives earnings across manufacturing, ecommerce and transport. In fact, every industrial sector has been transformed by technology with chips at its heart.

About 30 per cent of all chips still go into personal computers, some 20 per cent into smartphones, 10 per cent each into data centres and cars, with the remainder taken up for industrial and defence use, including tiny sensors installed as part of the 'internet of things' wave that will draw 'dumb' items into the 'smart' communications network of the future. Their constellations control and monitor and are heavily responsible for generating the world's store of data that conservatively doubles every two years.

Chips have scaled up or overshadowed the inventions that came before them: processing libraries of content, controlling cities of streetlights, improving crop yields and turning cars into computers on wheels as the internal combustion engine is readied for retirement. Their computational power handles tasks that humans alone could not fathom: cracking the code of life to combat disease, helping mankind conquer space – as well as less vital endeavours, such as mining bitcoin.

And there is more to come. Microchips continue to power the mobile revolution that freed devices from home, office and power supply. Compared to second- and third-generation (2G and 3G) wireless communications technology that underpinned most consumers' first cellphone, the coming fifth-generation (5G) standard promises to unleash high-resolution video games and mixed-reality viewing with the bandwidth to support permanent connections. Response times that are almost delay-free, taking place in the blink of an eye, give rise to confidence that robot surgery, autonomous driving and critical infrastructure can be managed remotely via communications networks.

It is a trend that the chip industry enables but will also feed off. More powerful networks mean more devices connected to them – and many of them will be more powerful devices, containing more advanced chips. Since their invention, microchips have been seen as a passport to prosperity, for governments eager to house high-value, strategic industries and for parents seeking to enrich their children's education with the latest personal computer.

Yet they also suffer from negative connotations, a symbol of invasive technology in dystopian fiction that controls thoughts or conducts Big Brother surveillance.

This point of view has seeped into real life. Witness the conspiracy theory that the Covid-19 vaccination programme was cover for a sinister plan to track the population in their everyday lives. When one survey of 1,500 American adults asked whether the US government was using the vaccine to microchip the population, 20 per cent of respondents said it was definitely or probably true.[18]

What cannot be argued is that for more than 60 years, chips have continuously improved, in price and processing power: a metronome of technological evolution. Consequently, their supply and the secrets required to make them efficiently are today worth more than the world's stock of gold or oil. They are already being fought over – verbally at first – and, because they are essential components in precision weaponry, are already fought with too.

Chip-powered progress is not without cost. Semiconductors are also great consumers, expanding to fill the political agenda, but also, when a new manufacturing plant costs $20bn to build, of large sums of capital.

That isn't all they expend. Harvard University research predicted that information and computing technology would consume 20 per cent of global energy demand by 2030, with most of that taken up by building hardware, notably microchips.[19] One-third of a mobile device's carbon dioxide emissions stem from manufacturing the chips it contains. In carrying out that process, a large fab can consume up to 10 million gallons of water every day through cooling and cleaning. Reducing chips' environmental footprint has become just as important as meeting the demands of every industry on the planet.

## Narrowing the Field

Imagine the maze-like tunnels of London's underground train network or the roads that crisscross a major city like Beijing. And then multiply them, billions of times, and shrink that grid of spaghetti junctions to something far smaller than the naked eye can see.

As an idea of how intricate a microchip is today it isn't bad, except that the latest chips are not simple street scenes, however tiny, they are high-rise tower blocks pulsing with activity. Each chip contains billions of transistors, the tiny switches that flick the flow of electrons on and off billions of times per second to run the computations that control the product they are embedded in or solve a problem. To pack more transistors in so each chip can do more, more quickly, latticeworks of circuitry are stacked as many as 150 layers high.

Each transistor, the building block of modern computation, is hewn from silicon, a perfect 'semiconductor', halfway between metal and non-metal, whose conductive properties can be dialled up or down depending on how they are mixed with other substances such as phosphorus or boron.

Chips smaller than a thumbnail feature structures as small as 3nm – that is, smaller than a biological protein, a fraction of the width of a human red blood cell, or a quarter of the size of the average virus. And the race is on to shrink further. More transistors on a chip means more computations can be carried out – increasing its power but at less relative cost.

In August 2021, Intel announced a breathtaking breakthrough: the first microchip to contain 100 billion transistors. 'Actually, I'm not even sure if it is accurate to call it a chip,' said Masooma Bhaiwala, the chief engineer of what was christened Ponte Vecchio, created to handle advanced artificial-intelligence tasks. 'It is a collection of chips that we call tiles that are woven together with high bandwidth interconnects that are made to function like one monolithic silicon.'[20]

Such was the relentlessness of the industry that few paused to consider this feat. In fact, seven months later, Intel had been trumped. Apple's new M1 Ultra chip, designed for use in its Mac Studio desktop computer system, boasted 114 billion transistors.

There is little chance the iPhone maker will stay out in front for long. ASML, a little-known Dutch company whose equipment is vital for pushing the boundaries of microelectronics, is plotting a path to over 300 billion transistors on one logic chip by 2030. Intel has one trillion transistors in mind in the same time frame. Meanwhile, the US artificial-intelligence company Cerebras Systems has already put 2.6 trillion transistors on a chip, although it was the size of a silicon wafer, the disc from which hundreds of square chips – also known as dies – are usually cut.

Engineers and customers alike are primarily interested in what these chips are capable of, rather than numerical bragging rights. Progress is shrinking a room-sized machine to a portable gadget; it is the iPhone, the 'everything device' that offers 100,000 times more processing power than the computer that guided Apollo 11 to the first moon landing in 1969. It is collapsing together numerous appliances – the phone, camera, calculator and games console – because it can, because the technology meets the vision.

---

Making microchips is a global endeavour. The sand from which silicon is extracted could be dug from one of several leading mines such as the Quartz Corporation's Spruce Pine facility in North Carolina and then shipped to the village of Drag in northern Norway for refinement – the first sign that high-value, light-weight goods mean distance is no object. From there, the rocks could go to Japan, where materials firms specialise in making silicon wafers, the thin, rainbow discs onto which chip designs are printed.

In the vast factories, or fabs, mainly based in Asia, the wafers are whisked noiselessly from process to process in robot pods that travel along ceiling-mounted conveyor belts. They clock up thousands more miles without leaving the building as they undergo 3,000 processes over a three-month spell. In these cleanrooms, into which purified air is pumped to ensure the atmosphere is pristine with 1,000 times fewer dust particles than in a hospital operating theatre, technicians dressed in head-to-toe white 'bunny suits' look on.

It is also a hugely expensive business. TSMC spent the best part of $20bn to build Fab 18 in Tainan, southern Taiwan, which went into production in 2020 with floorspace of 950,000 square metres (over 10 million square feet) – the same as 133 football pitches.

Inside, wafers are coated with layers of light-sensitive materials and repeatedly exposed to patterned light through a photomask, or stencil, which itself costs millions of dollars to make. The parts of silicon not touched by the light are then chemically etched away to reveal, gradually, the intricate details of a chip that are built up, layer after layer. The cleanroom is bathed in yellow to filter out light at short wavelengths that would cause an adverse reaction during the production process.

When complete, each wafer disc is laser-cut into perhaps thousands of chips, carefully packaged, tested and sent to the customer's own production line for installation and, ultimately, into the hands of the consumer.

The complexity of the manufacturing process is matched, detail for detail, by the design process. Developing chips can take several years, thousands of engineers and millions of dollars. Just like building a house, there are architects and floorplans to pore over, plus functionality, cost and speed to consider. With fabrication so costly, so much must be simulated first to avoid expensive mistakes later.

Taken together, an industry that was born in the 1960s with many companies doing many things has splintered into a series of specialists. The search for better processes and reality of great cost has narrowed each field within design, manufacture, manufacturing equipment and packaging so that one or two players dominate. As demand for more – and more capable – chips rises over time, fewer companies are capable of fulfilling the world's requirements because a small handful have performed awesomely well – often over decades spent perfecting just one aspect.

No wonder TSMC is in the political spotlight, but it is not alone. To make the most advanced chips, there is currently no alternative to the extreme ultraviolet (EUV) lithography machines produced by the Dutch company ASML. They cost €160m a time, are the size of a small bus and took 20 years to develop.

There are other strongholds, and one notably in intellectual property, which provides this book with its core narrative. Almost equidistant from the American microchip titans and the Asia manufacturing powerhouses, at the industry's fulcrum lies a company that began life in Cambridge, the UK's leading university city. In the traditional sense it makes nothing; it has no laser machinery nor cavernous factories to speak of. But, as it quietly gains market share year after year, it feels as though its ideas supply everything.

## A Valuable Rulebook

Advanced RISC Machines, or simply Arm these days, is far from being a household name. Yet its designs have infiltrated billions of households, workplaces and vehicles the world over since it was founded in 1990.

Just as chipmakers ceded manufacturing to contractors as it became more complicated, they began to buy in ideas for the design process too. Providing pre-set patterns, these blocks of intellectual property offered another shortcut as cost and complexity ratcheted up. Arm has been a key beneficiary, earning its spurs by powering the mobile-phone revolution that began 25 years ago.

The company owns a valuable rulebook, or instruction set architecture (ISA), for chip design. Used the world over to determine how a chip's central processing unit (CPU) – the 'brains' of the device – is controlled by software, it is found today in numerous mobile phones, cars, laptops, data centres, industrial sensors and more.

The ISA is a kind of digital-era Ten Commandments, offering the predictability that helps computer developers to write more efficient code by defining what the machine will do but not how it will do it. Wherever a piece of software is used, any Arm-based processor carries out instructions in the same way.

Arm's use case has grown long but its ISA is anything but. It consists of a few thousand instructions – or rules – but they can be configured into four billion possible encodings. Amazingly for something so embedded in digital life, its relevant materials are still available in hard copy. Arm's Architecture Reference Manual (the Arm Arm) sets out over 10,000 pages how to use the ISA as well as offering troubleshooting advice, written in a language that looks a lot like C, one of the most popular computer programming languages.

The ISA is also a living document, updated quarterly by a team of 40 people in Cambridge, sometimes by adding new functionality, such as enabling more major multiplication that is needed for machine learning, adding more security features, addressing customers' problems or correcting glitches. There are major overhauls that are typically carried out once a decade, most recently in 2021.

In explaining Arm's reach, it is worth segmenting the chip industry. Of the more than one trillion chips sold every year, the vast majority do not require the type of processor that Arm designs. Broadly, memory chips that store information are simpler and more commodified. Most analogue, optical and mechanical chips fall outside the company's scope too. Where Arm plays is in the logic space, where chips act as the 'brains' of electronic goods that process information, and includes microcontrollers as well as communications chips.

Depending on the licence they buy from Arm, developers can use its designs off-the-shelf, build on them slightly or greatly vary them, while still maintaining compatibility. Arm helps people to use its designs. For example, it offers a compliance kit, which is a set of tests to check that code written by licensees is performing properly, overseen by 100 engineers based in Bangalore, India.

One prong of Arm's success derives from the fact that chips with the greatest capacity are not always the best for the task at hand. Performance must be set against power consumption – both at the peak of a chip's operation and when it is sitting idle – as well as the amount of silicon required, because that relates to cost.

These variables have guided how the industry developed, and why Arm's low-power and low-cost designs have prospered. So has Moore's Law, the prediction that the number of transistors that could be fitted on a chip would double every two years – of which more later.

Arm has succeeded by walking with giants. FutureHorizons estimates that the biggest tech firms, including Apple, Google and Amazon, which have focused in the last decade on designing their own chips and cutting out the middlemen, such as Intel, earn the lion's share of industry revenues. That equates to an average of $450 per square centimetre of silicon used, dwarfing the foundries such as TSMC, which can expect to pick up $4. The intellectual property provider, often Arm, can expect only 10 cents per square centimetre.[21]

It doesn't sound much. And it's true, Arm's designs cost relatively little and are used often: an astonishing 29.2 billion times in 2021, around 60 times as frequently as Intel, long seen as the microchip industry leader. Having doubled volumes in six years, Arm's is the most widely deployed computer hardware on the planet – more than any PC type, more than any smartphone.

And it is still growing. Some 13 million software engineers, greater than the population of many countries, write code that runs on Arm. A prime mover in the mobile revolution, it stands to gain much from the spread of 5G. As chips become brainier, guiding and sometimes dictating a greater part of our lives, running more and more software, there is a fair chance they will do so containing Arm.

This design, these conventions, have rapidly become the 'everything' blueprint, a global technology standard at the heart of modern computing and consumer electronics, that has changed the world.

How Arm got here is a remarkable story. But first, before Arm got its chance, before chip manufacturing migrated to Asia, it's important to understand the industry's all-American origins.

## Chapter 2

# SOME HISTORY: THE ODD COUPLE STARTS OUT

### The World Awaits

It was the buzz about the Institute of Radio Engineers' (IRE) 49th annual convention and an associated flurry of advertising that swelled the 10 March 1961 edition of *Electronics* magazine to a thumping 316 pages. The American industry bible predicted that more than 70,000 engineers would descend on New York's Waldorf-Astoria Hotel and the nearby Coliseum conference centre later that month to share their vision for the future.

Over four days, they would shuttle between two venues alongside Central Park to hear speakers that had jetted in from as far away as Norway, Japan and Venezuela deliver 265 research papers, as well as peruse more than 850 exhibits.[1]

'On the Coliseum's 4 gigantic floors you'll see the latest production items, systems, instruments and components in radio-electronics; in radar; in complex air traffic control; in space communications – in any and every field of radio-engineering you care to name,' an advertisement for the event breathlessly announced.[2]

If the 1950s were for dreaming, as the Western world shook off war-induced austerity, there were new realities as the 1960s dawned. Some 30 years before Arm was founded, the Cold War space race was on, as the US and Soviet Union attempted to outdo each other's satellite technology. Mobile gadgets, smart networks and artificial intelligence were still a science-fiction dream, but rising incomes and slick marketing campaigns meant the latest TVs and white goods kitted out homes.

In January 1961, John F. Kennedy's presidential inauguration was the first to be broadcast in colour. His predecessor, Dwight Eisenhower, had enjoyed a landslide victory accurately predicted soon after polls closed in 1952 by the Universal Automatic Computer (UNIVAC) for the CBS TV network. The hulking, grey unit, weighing over seven tons, offered most viewers their first glimpse of a computer.

Enchanted by these new technologies, political leaders and consumers thought the possibilities endless. The trouble was the engineers behind such radical advances knew precisely the strictures within which they were operating.

More elaborate functions demanded gadgets stuffed with more elaborate circuitry. But what was being dreamed up in the research laboratory could not be made real on the production line. It was simply too complicated, costly and time-consuming to manufacture. Every transistor that amplified or switched on and off an electronic signal needed connecting to thousands more components – resistors that reduced the flow of current, capacitors that stored and released energy and one-way diode switches – to create a circuit in a continuous loop. The only way to do that was by wiring each element together by hand.

'It was almost entirely women's work, because male hands were considered too big, too clumsy, and too expensive for such intricate and time-consuming tasks,' the author T.R. Reid wrote.[3] There was a stench of sexism about it too. Women had programmed wartime computers but a generation on there were still few technical or managerial jobs available to them. Beneath magnifying glasses, they wielded tiny soldering tools and tweezers. Mistakes were unavoidable.

'For some time now, electronic man has known how "in principle" to extend greatly his visual, tactile, and mental abilities through the digital transmission and processing of all kinds of information,' Jack Morton, a vice president at Bell Laboratories, the US research centre established by the inventor of the telephone Alexander Graham Bell, wrote in 1958. 'However, all these functions suffer from what has been called "the tyranny of numbers". Such systems, because of their complex digital nature, require hundreds, thousands, and sometimes tens of thousands of electron devices.'[4]

That tyranny persisted in 1961. The legion of engineers flocking to the Waldorf-Astoria were the ones determined to overthrow it and *Electronics* magazine was eager to provide an indispensable guide to the bumper IRE show. In a highlights feature, writers picked out what they thought would be new and interesting for delegates, including radar, log-periodic antennae and tuneable tunnel-diode amplifiers. However, one innovation that eluded the title's previewers was something that would prove in time to be a gamechanger.

'Announcing the first of a new family,' the company Fairchild Semiconductor proclaimed that week in its product brochure: 'The Micrologic Flip-Flop'. This basic electronic storage circuit for a single bit of data that flipped between two steady states of one and zero, was 'the first element of the micrologic family of digital functional blocks'. It would be followed soon by five other devices that together were 'sufficient to efficiently build the complete logic section of a digital computer or control system'.[5] Incidentally, a 'flop' was short for 'floating point operations per second', another measure of performance related to the type of arithmetic required to tackle very large or very small numbers.

Simpler to understand was that the scientists behind the Micrologic appeared to have worked out how to do away with the fiddly wiring. Theirs was the first commercial 'integrated circuit' (IC) that featured all the required components on a single piece of silicon. The microchip was born.

However, the excitement Fairchild generated did not translate into sales. At $120 a time, its new product was comfortably more expensive than a circuit wired together by hand. Not for the last time, a solution to the electronics industry's biggest problem had presented itself but was not yet judged to be viable.

As well as winning over customers, Fairchild had to be mindful of the competition. In fact, what looked very like an IC had debuted at the same event two years earlier when one of the biggest names in the industry, Texas Instruments, showed off what it called a 'solid circuit'. The technological race was on.

## The Unassuming Kilby

The fathers of the microchip are an odd couple, an introvert and an extrovert whose companies' decade-long patent battle conceded joint parentage and a cross-licensing agreement. Never achieving the fame of inventors Henry Ford or Thomas Edison, Robert Noyce and Jack Kilby are forever yoked together thanks to an invention that has become far more pervasive than either the motor car or the light bulb.

Born in 1923, Kilby was tall, quiet, unassuming, well-read, a fan of big-band music who did his best work alone. Four years his junior, Noyce was almost the opposite: a collegiate and charismatic leader, an action man who in later life piloted his own plane to meetings. He skied, scuba-dived and sailed with aplomb and in time evolved into a slick media performer on behalf of his industry.

Kilby had his father to thank for an early radio enthusiasm that grew into a fascination for electronics. Kilby senior ran a power supplier in Kansas and when a huge ice storm felled telephone and power lines, he borrowed a neighbour's radio to keep in touch with customers. His curious son soon built his own amateur set so he could listen into the night.

After falling short in the maths entry exam for the prestigious Massachusetts Institute of Technology (MIT), Kilby studied electrical engineering at his parents' alma mater, the University of Illinois. In his first job, at an electronics manufacturer in Milwaukee, Wisconsin, that made parts for radios, televisions and hearing aids, Kilby had the twin tenets of cost and reliability drilled into him during the day, while studying for a master's degree in electrical engineering at night. He was a problem solver. For his industry to prosper, Kilby could see that size mattered.

In 1958, he moved with his wife to Dallas, Texas, to join Texas Instruments (TI), 'the only company that agreed to let me work on electronic component miniaturization more or less full time, and it turned out to be a great fit'.[6]

From its roots as a geophysical research firm whose sound waves led the hunt for oil deposits, TI was changing direction and staffing up. Its go-getting president Patrick Haggerty had already found the scientists to develop a cheap transistor that could be mass-produced and hit upon the device that drove demand for electronics from millions of consumers: the pocket radio. Now Haggerty was eyeing a bigger challenge – how to solve the tyranny of numbers.

———

That puzzle only presented itself because of another invention just over a decade earlier. The transistor recast the potential of the electronics industry when it elbowed aside the vacuum tube, which had powered devices for the first half of the century via wire filament that carried electric current inside a sealed glass bulb. The UNIVAC computer used 5,000 vacuum tubes; the ENIAC, a predecessor model, included 18,000. Their output might have wowed the crowds but the tubes that drove them were hot, fragile and huge.

At Bell Labs in New Jersey in 1947, the physicist William Shockley led the team that discovered the transistor – although the scrap over who did what towards that breakthrough surely contributed to his exit several years later. Regardless, it was a stunning scientific advance: out went the vacuum tube, in came 'solid state' semiconductors to take control of gadgets from here.

Bell Labs was part of AT&T, the monopoly US telephone company. Encouraged by the government, it realised that the development of the transistor would move faster if word spread. Bell Labs organised a symposium in September 1951 to explain the transistor's potential to 300 scientists and engineers. They left the event excited, but none the wiser about how to build one. For that, they had to buy a $25,000 licence and gather again the following April. The information shared at that meeting was poured into a two-volume book set called *Transistor Technology*, which soon earned the nickname 'Mother Bell's Cookbook'.

Just like the tiny components that would in time be packed on a single piece of silicon, the transistor's invention was really the sum of its parts, a roll call of brilliant people – largely men – who laid the groundwork for what would follow.

Shockley owed a debt to John Bardeen and Walter Brattain, with whom he would share the Nobel Prize for physics. All three must nod to the Austro-Hungarian physicist Julius Lilienfeld, who proposed the concept of a field-effect transistor in 1925 but was not able to construct a working model, and many other scientists that had gone before.

Shockley was in the right place. Since 4,000 scientists and engineers had been assigned to the newly created Bell Telephone

Laboratories in 1925, it had become a premier research institution, making great strides in talking movies, radio astronomy,
solar cells, calculators and cryptography. And it did not take long
for the brightest minds to build on Bell Labs' latest invention,
the transistor.

In 1952, Geoffrey Dummer, a manager at the UK's Telecommunications Research Establishment, a development unit that had
worked closely with the Royal Air Force on radio navigation, radar
and infra-red detection for heat-seeking missiles during the Second
World War, noted that given the advent of the transistor and work
in semiconductors generally, 'it seems now possible to envisage
electronic equipment in a solid block with no connecting wires'.[7]

Sure enough, within months of arriving at TI, Kilby had something to add himself. The newcomer had not earned any vacation
time so was left alone to think in the lab over summer. Those
undistracted hours led him to ponder whether all the parts of
a circuit could be made out of the same material and therefore
integrated onto the same base. That way, connections could be
printed out instead of wired and space would be freed up to cram
on more components.

Using a small piece of the grey semiconducting metalloid germanium, by early 1959 Kilby's first crude demonstration chip,
half the size of a paper clip, was ready. Because he used an older
transistor, capacitor and resistors, some of them stuck up over
the plane of the semiconductor and he had to make a few connections by hand with gold wire. It was messy, but it was a start.

## The Adventurous Noyce

When Robert Noyce answered the phone to William Shockley in
January 1956, it was a thrilling, out-of-the-blue conversation for
the young Iowa-born physicist. 'It was like picking up the phone
and talking to God,' he said later of the request for him to come

to California for a job interview. 'He was absolutely the most important person in semiconductor electronics. Getting that job meant you would definitely be playing in the big leagues.'[8]

Shockley needed no introduction. After his famous break-through, he was intent on commercialising the transistor and wanted to hire promising young talent to join his new company. The ambitious Noyce had caught his eye thanks to a recent research paper he had delivered.

The son of a preacher, Noyce's youth spent in the open expanses of the US Corn Belt was enlivened by escapades that involved building radio-controlled airplanes, pig rustling and launching himself off a barn roof while clutching a homemade glider. His desire to understand how things worked steered him towards degrees in physics and maths at his local Grinnell College. At the Massachusetts Institute of Technology (MIT), where he graduated with a PhD in physics, he was nicknamed 'Rapid Robert' for his quick thinking and stood out from the crowd with film-star good looks.

The call from Shockley was well-timed. Following his studies, Noyce had spent a few underwhelming years as a research engineer at the electronics firm Philco in Philadelphia, where he had grown frustrated with corporate life and was ready for a change.

Shockley Semiconductor Laboratory was to be based in Palo Alto, California, because Shockley was intent on moving back west from New Jersey to be close to his mother. Since he grew up there, the area had transformed itself into a hub for radio companies that prospered during the war years, bolstered by a steady flow of graduates from the West Coast universities.

Compared to Harvard and Yale out east, where professors worked full-time, Stanford and, 40 miles north on the other side of San Francisco Bay, Berkeley, were less buttoned-down. That culture could be traced to Frederick Terman, an electrical engineer who worked on vacuum tubes and circuits and returned to become dean of Stanford's school of engineering after the Second World War.

Terman let his academics loose to do other things one or two days a week.[9] In this liberal spirit of capitalism, ideas could be socialised and commercialised and companies were formed and often set up on Stanford grounds. Risk-taking was encouraged. Early tenants of Stanford Industrial Park, that lay adjacent to the university and was later renamed Stanford Research Park, were Terman's former students William Hewlett and David Packard, whose audio oscillator, manufactured at his encouragement, was bought by the Walt Disney Corporation.

Hewlett-Packard became a mainstay of Silicon Valley – although the area wouldn't be christened that by an enterprising journalist until 1971. To have earned the title, it was Shockley seeding talent in the area that was pivotal.

One of those new arrivals, Noyce, was so confident of landing his new job that he put down a deposit on a house close to the company's base several hours before his interview took place.[10] He was duly appointed, but the opportunity was short-lived.

The brilliant team of twenty-somethings Shockley had brought together were excited when their boss was awarded the Nobel Prize in physics. But even before the jubilation faded, it was clear they were working for a tyrant who had little time for them or their input into product development. Seven of the firm's original recruits resolved to quit. They needed a ringleader, and Noyce, who had been elevated to become a favoured manager, agreed to join them. The 'Traitorous Eight' – as the irate Shockley labelled them – served their notice in September 1957, gaining industry renown.

'We didn't realize at the time the legacy we'd leave,' said Jay Last, another of the eight. 'Thank God Shockley was so paranoid or we'd still be sitting there.'[11] At least his academic credentials were unimpeachable. Later in life, they would be overshadowed by Shockley's controversial eugenics work, including a proposal that people with low IQs should be paid to undergo sterilisation.

––––––––––

The breakaway team might have tried to find another employer until an enterprising investment broker, Arthur Rock, introduced to them by a family connection, suggested they look for a backer to help them set up on their own. If it hadn't been for Sherman Fairchild, Rock suspected the eight would have gone their separate ways – or gone to Texas Instruments.[12]

Fairchild grew up wealthy enough to take risks safe in the knowledge they didn't all have to pay off. He was an only child and sole heir to the fortune of his father, who had been the first chairman of International Business Machines (IBM), which started out selling punch-card record-keeping equipment. That left fun-loving Fairchild the funds to wine, dine and invent things, including aerial cameras that were used by the US Air Force in the war. Keen to expand into the hot area of transistors, Fairchild Camera and Instrument Company stumped up $1.5m to back the Traitorous Eight and, in late 1957, Fairchild Semiconductor was born.

That was the easy part. There was no off-the-shelf technology of which to take advantage. From its 1,300 square metre (14,000 square foot) base in Palo Alto, Fairchild made its own silicon wafers and built its own furnaces for diffusion, the process by which the semiconductor's electrical properties were modified when impurities were introduced and baked at high temperatures. But Noyce's charm drew in early contracts that no tiny start-up could have realistically hoped for.

Fairchild's great technological stride forward began in 1958 when Jean Hoerni, a Swiss engineer and one of the Traitorous

Eight, was trying to improve the process for the transistors that would supply the US government's Minuteman ballistic missiles. He proposed to protect them with an insulating layer of silicon dioxide to improve reliability, reduce susceptibility to contamination and enable high-volume production in what became known as the 'planar' process. In doing so, Hoerni made the device much flatter. Noyce took things a step further, adding a conducting metal pattern across the top to connect the transistors together without the need for wiring.

The initial burst of interest in the integrated circuit at the Institute of Radio Engineers' annual convention in 1961 did not convert immediately into sales. In fact, some inside Fairchild wanted to focus efforts on their popular and highly profitable transistors and discontinue ICs altogether. As tensions flared between founders, Hoerni had already quit to set up Amelco, an IC firm that was the first of numerous 'Fairchildren' ventures that would share the same heritage.

The US government was the technology's saviour. The second-generation Minuteman that went into production in 1966 was the first high-volume use for integrated circuits. The missiles could strike targets from a greater distance and with greater precision than the first-generation weapon and stood poised for action during the Cold War.

In addition, the US President, John F. Kennedy, was eager to beat the Soviet Union in the space race. Having watched Sputnik 1 launch in October 1957, the National Aeronautics and Space Administration (NASA) and the US Department of Defense became big IC customers. They were unfazed by pricing, and 2 per cent of the country's gross domestic product (GDP) was ploughed into research and development. Offering vastly improved processing power, these relatively small and lightweight devices found a home in the Apollo Guidance Computer (AGC) installed on board each of the command modules in the space programme that led up to the 1969 moon landing.

State support didn't end there. While inclusion in the AGC stimulated interest from other industries, a 1977 study showed that the US government provided just under half of all research and development money spent by the US electronics industry in the first 16 years of the microchip's life. The government was responsible for all sales until 1964 and remained a significant buyer thereafter.[13]

---

Even then, the potential of the IC was difficult to comprehend. One man who thought he could envisage the direction of travel was Gordon Moore. Another one of the Traitorous Eight, Moore was director of research and development (R&D) at Fairchild Semiconductor when he was asked by *Electronics* magazine to predict what would happen in his industry over the next decade. The resulting article, 'Cramming more components onto integrated circuits', published on 19 April 1965, outlined what became known as Moore's Law, extrapolating that computing would dramatically increase in power and decrease in relative cost at an exponential pace.

Its premise seemed preposterous. Given the trend that Moore had observed over the last few years, he thought that the number of electronic components – including transistors, resistors and capacitors – that could be squeezed onto a microchip would double every year for the next decade. In 1965, Fairchild was preparing to deliver chips containing 64 separate components to a handful of customers. Moore's prediction meant that by 1975, the number would total 65,000. 'I believe that such a large circuit can be built on a single wafer,' Moore wrote. In addition, he foresaw many of the end markets for integrated circuits, including 'home computers', 'automatic controls for automobiles' and 'personal portable communications equipment'.[14]

By 1975, Moore thought the march of miniaturisation had further to go. Looking forward another decade, he revised the forecast to a doubling every two years, never thinking his yardstick – which was never really a law – would endure for so long and give enthusiastic scientists something to aim for.

'Rather than becoming something that chronicled the progress of the industry, Moore's Law became something that drove it,' Moore said years later.[15] His name would loom larger and longer over the industry than that of his colleague Noyce the showman ever would.

In fact, the calm, considered Moore was the antithesis of Noyce: he sweated the details rather than skated on top of them. The fifth-generation Californian earned a bachelor's degree in chemistry from the University of California at Berkeley and a doctorate in chemistry and physics from the California Institute of Technology. He spent his downtime fishing, a hobby as meditative as the way he worked, which had won him a strong following inside Fairchild. Aside from his bold predictions, it made Moore an essential recruit when Noyce once again prepared to jump ship.

## A New Era

The early calculator wars offered a template for consumer electronics battles to come. Compared to the first all-transistor calculator, IBM's 608, which was housed in several large cabinets and retailed for $83,000 when it went on sale in 1954[16], devices shrank in size and price as processing power multiplied and manufacturers could take advantage of economies of scale. But as gadgets that consumers didn't know they needed quickly became must-have status symbols, the stampede of competitors meant profit was harder to come by.

By 1969, Busicom was one of the also-rans. The Japanese electronics firm was faring badly in an overpopulated market and reasoned it had nothing to lose. Searching for a great leap forward in calculators, it turned to Robert Noyce, who was venerated by Japanese engineers for his invention of the integrated circuit.

Noyce needed the business too. At Fairchild Semiconductor he had grown restless and felt that the parent company Fairchild Camera and Instrument wasn't reinvesting enough of the proceeds from his highly profitable semiconductor business into research and development.

As he conveyed in his resignation letter, Noyce wrote: 'I do not expect to join any company which is simply a manufacturer of semiconductors. I would rather try to find some small company which is trying to develop some product or technology which no one has yet done. To stay independent (and small) I might form a new company, after a vacation.'[17]

Noyce and Moore incorporated their new venture, NM Electronics, on 18 July 1968, but quickly renamed the company Intel, a portmanteau of 'integrated electronics' that Noyce thought 'sounded sort of sexy'. Following them over from Fairchild was Andy Grove, a young physicist who had worked under Moore in R&D. There was financial backing once again arranged by Arthur Rock, who became Intel's chairman. 'I was never as sure that a company would succeed as I was that Intel would,' he said.[18]

The new company favoured memory chips over logic because they were easier to design than logic chips, but after a few tough years it went where the work was. Busicom commissioned Intel to manufacture a set of a dozen specialised chips it had designed for its next-generation calculator that combined the key functions of memory, logic and input/output that communicated with the outside world. The project was handed to Ted Hoff, who went a step further by condensing Busicom's requirements onto four chips, including just one that contained the logic circuitry of the device's central processing unit (CPU).

Its architecture was refined and designed into silicon by a new Intel hire, Federico Faggin, who had come to California from his native Italy after joining Fairchild's joint venture there.

The great innovation behind this CPU-on-a-chip – soon to be labelled a microprocessor – was to abandon cumbersome customisation in favour of creating a general-purpose chip that could be mass produced and programmed with software to perform specific tasks . . . in this case, numerical calculation.

It was a significant breakthrough, but the bottom was falling out of the desktop calculator market while it was being made. Busicom was keen to renegotiate its agreement so Noyce took the rights to use the chip for everything but calculators and returned $60,000 of development money to the Japanese. By 1974, Busicom was bust, but Intel was on to something.

'Announcing a new era in integrated electronics,' an advertisement in *Electronic News* proclaimed in November 1971. For $60 a time, the 4004 was 'a micro-programmable computer on a chip!'[19] And so it began.

# PART ONE

# ARM (1985-2000)

## Chapter 3

# FROM A TINY ACORN, DESIGNS ON THE FUTURE

## A Comical Confrontation

On the last Friday before Christmas 1984, as office workers and academics unwound into the festive season, Sir Clive Sinclair pushed his way through the crowd at the Baron of Beef pub in the centre of Cambridge clutching a rolled-up newspaper.

The dome-headed electronics entrepreneur, sporting a ginger beard and spectacles, was instantly recognisable thanks to regular TV appearances and a star turn in his company's own advertising campaign. And he was fuming.

Across the room Sir Clive saw the object of his ire, his blunt, intense former lieutenant turned arch competitor, Chris Curry, who sported generous dark sideburns and a suit and tie. Together the pair had worked to launch all manner of electronic devices: amplifiers, calculators, pocket radios, miniature TVs and watches. But the gadget of the moment, the home computer, saw them pitted against each other in a market that was rapidly overheating.

The source of Sir Clive's rage was a national newspaper advertisement, placed by Curry, that questioned the reliability of Sinclair Research's ZX Spectrum computer. The full-page display implied that if buyers did not want to have to return their device after Christmas, better to give an Acorn Electron or BBC Micro in the first place. Both were produced by Acorn Computer, the venture Curry had co-founded after leaving Sir Clive's side.

Sir Clive saw red. That night he let rip an expletive-laden attack on Curry, bashing him with the newspaper and entertaining revellers. 'He was extremely aggressive and rude to me and he was calling me names,' Curry said in one report of the fracas. 'I tried to placate him but it was no good.'[1]

It was a comical confrontation but what lay beneath was deadly serious. Even though Curry claimed Apple, the American computer maker run by the free-spirited Steve Jobs, was his biggest rival, Acorn's campaign knocking Sinclair was a desperate, last-ditch attempt to ignite sales. Both sides' finances were stretched to the limit because they had agreed on one thing: Christmas 1984 was going to be another bonanza for home computer sales. They were wrong.

As Sir Clive pursued Curry to a nearby wine bar to continue the row, both knew that 1985 would bring with it some hard truths. Amid tumbling prices and a flood of supply, their leadership of a sector they had done much to shift from hobbyist computer builders to casual computer users was slipping through their fingers.

The bust was coming after a remarkable boom, which was sparked three years earlier by a unique intervention.

## The Computer Programme

A little after three o'clock on the afternoon of 11 January 1982, Britain's TV-viewing public was greeted with a glimpse of what many must have felt was a far-off future. The first episode of *The Computer Programme* on BBC2 featured a Cray-1 supercomputer, stood like an imposing wardrobe, processing 50 million instructions a second to produce a 10-day European weather forecast. Less abstract for the layman was a short report on Phyllis, a veteran sweet-shop owner and unlikely early technology adopter, enthusiastically keying the day's stock take into her home computer after hours.

'One thing I know already: don't expect the computer revolution to happen tomorrow,' declared the programme's presenter Chris Serle, who, dressed in brown jacket and tie, resembled a kindly college lecturer. 'It's happening now.'[2]

The 10-part series was meant to do more than demystify this new technology for the curious masses. It was the cornerstone of an ambitious, government-supported plan to encourage computer adoption in homes, schools and small businesses across the land.

The Computer Literacy Project (CLP) had its roots in the BBC's Continuing Education Television Department, which championed adult self-improvement. Several popular documentaries had aired on the subject, but executives thought there was more to do. Leaning on the expertise of several government departments, the BBC despatched two journalists, David Allen and Robert Albury. Their fact-finding mission to France, Holland, Germany, Sweden, Norway, Japan and the US, funded by the Manpower Services Commission, a public body with a remit to co-ordinate UK employment and training, yielded a three-part TV series, *The Silicon Factor*, and 'Microelectronics', a neatly typed, 50-page document published in December 1979 and dropped onto the desks of every UK Member of Parliament the following summer.

Ranging across video discs, voice synthesis, electronic mail and automated factories, Allen and Albury declared they had obtained a 'broad and non-parochial view of the way in which the new technology is being applied' around the world. Their report's aim was to shape the UK's response to mainstream computing by 'helping define promising areas of multi-media educational provision'.[3] It was a challenge that echoed down the decades: how developed nations ensure their people are equipped to thrive in the face of the latest scientific advances.

---

Certainly, it was not the first time the UK had wrestled with this threat before, notably in the future prime minister Harold Wilson's 'white heat' speech given in 1963, almost a generation earlier, when he was still in opposition.

Wilson, a pipe-smoking former economic history lecturer, called for the creation of more scientists and the application of their efforts more purposefully to boost national production. If backs were turned on wholesale industrial changes such as automation, 'the only result will be that Britain will become a stagnant backwater, pitied and condemned by the rest of the world', he warned.[4]

In government, Wilson practised what he preached, in 1968 overseeing the three-way merger that created International Computers Limited (ICL), which was intended as a British champion capable of taking on major manufacturers such as America's IBM.

One of the points the 'Microelectronics' document raised for discussion was that this new technology needed democratising for both children and adults. 'People who are curious or anxious must be helped to understand the nature and the effects of new technology,' it said. 'It needs demystifying. The black box needs to become a grey box. It must not be thought to be understood only by an elite, let alone controlled by it.'[5]

Commissioned in November 1979 with the provisional title 'Hands on Micros', the BBC's next TV series on computing took on fresh importance. Producers were focused on how the project could have maximum impact and burst forth from the screen as an entry point to practical computing. In the great tradition of the

BBC's founder, Lord Reith, to inform, educate and entertain, it was not unusual for education programmes to be accompanied by additional learning materials supplied to schools and colleges. But this time the state broadcaster went significantly further.

Despite there being around a hundred different computers on the market when the TV programme went out, the BBC took the radical decision that it wanted its own machine on which to guide viewers through on-air demonstrations. The problem was the Basic computer programming language, which was widely used but had no common standard. The manufacturers approached by the BBC 'couldn't agree' on coming up with one, according to Allen, who became the CLP's project editor. 'So we decided that we needed one of our own which we thought was better than all of theirs.'[6]

It explains why, when Serle loaded a simple computer game from a cassette tape on that first episode of *The Computer Programme*, he did so on a BBC Micro.

With only three national TV channels to choose from, a generation of British children made for a captive audience. For the company that made the device for the broadcaster, it was a gold-plated commercial opportunity.

## Cambridge Brains

Acorn Computer had already achieved some success with the Atom, a computer sold for £120 in kit form so hobbyists could assemble it themselves, or ready-made for £170. The green lettering on the back of the white plastic-moulded box spelled out its birthplace: Acorn Computer Cambridge England.

Born and raised in the city, one of Acorn's two founders, Chris Curry, expertly marketed the Atom explicitly for home use when it was launched in 1980. He handled the commercial side of things while his business partner, Hermann Hauser, looked after most technical matters. To that end, recognising how important Cambridge University brainpower was to Acorn's continued success, Hauser identified an unlikely secret weapon: Fitzbillies' cakes.

To ensure Acorn had access to a steady stream of bright young academics, Hauser resolved to feed them. He let it be known that at 4pm most days there was tea and cake for anyone that passed by the firm's tiny premises, which were down an alleyway off Market Hill and up a flight of stairs, above the Eastern Electricity Board. Based a short walk away, many members of the university's computer laboratory habitually showed up for a sticky Chelsea bun from Cambridge's best-known bakery and a chat that often segued into a design meeting over dinner.

It was exactly this kind of studied informality that attracted Austrian-born Hauser to Cambridge in the first place. He had fallen in love with its ancient lanes and scholarly air aged 16, when his father, a wine seller, sent him there to learn English. The tall, elegant, blond-haired Hauser had returned to the city every summer, working as a research assistant in the Cavendish laboratory, Cambridge's renowned physics centre where academic giants including Ernest Rutherford had advanced works on radioactivity and J.J. Thomson discovered the electron. When it came to decide where to study for his own physics PhD after graduating from Vienna University, there was no contest.

Curry's background differed. While at school he sourced TV components from rubbish dumps to make amplifiers for local rock bands and did not go to university, intent on making some money instead. A few months at the start of his career were spent working for Pye, Cambridge's most famous electronics firm, whose brand was carried on TV and radio sets.

In 1966, Curry joined Sinclair Radionics and over the next decade grew to become a vital aide to Clive Sinclair, a journalist turned inventor whose minimalist design and relentless work ethic meant he could usually make a splash in emerging electronics categories – so long as his advertising campaigns hit the mark. However, the firm's digital wristwatch could not emulate the great success of an earlier pocket calculator, so the firm was part-nationalised by the government's National Enterprise Board (NEB) in 1976 after losses mounted.

When the involvement of the NEB became too much, Sinclair despatched Curry to set up another firm, Science of Cambridge, that he eventually intended to join. It had a mail-order hit selling a basic computer kit, the MK14 – so called because it had 14 components – comprising a simple circuit board with a small keypad and calculator display. Curry wanted to develop the MK14 further, but Sinclair did not, so Curry pursued his interest in microprocessors with Hauser, who he knew socially. In 1978 they set up together a consultancy business, Cambridge Processor Unit (CPU), initially sharing premises with Science of Cambridge above a parade of shops.

'It was a very gentle, almost unnoticeable separation,' Curry recalled. 'It was much later that it became a bit more competitive.'[7]

---

CPU needed staff. It was for this reason that one day Hauser invited himself around to Steve Furber's office, in an old house across the road from the university's engineering department.

Having diverted from studying maths into aerodynamics, the unassuming Furber was a Rolls-Royce research fellow under the celebrated Professor Shon Ffowcs Williams, whose own work involved efforts to make Concorde quieter. Furber was also a guitar player who had built a wah-wah pedal and a sound mixer. Thinking he could put together his own flight simulator one day, he made printed circuit boards, dousing them with ferric chloride in the sink to etch the design, and ordered microchips from

California to build a computer on which he had written his PhD thesis.

Hauser had spotted Furber at the Cambridge University Processor Group (CUPG), where like-minded computer enthusiasts gathered, and invited him to get involved with the new business. Making clear electronics was just his hobby, 'If he was interested then I was interested in tinkering,' Furber said. No money changed hands. Furber would create electronic designs as required and Hauser would keep him supplied with components to feed his pastime.

CPU's first project was to produce electronic controllers for fruit machines. The Welsh client got in touch because of the success of the MK14, as did Roger Wilson, another CUPG member, who was transgender and later transitioned to become Sophie.

Wilson was fiercely intelligent and good with their hands, brought up in a remote part of North Yorkshire by teacher parents who built their own family car and boat and made soft furnishings for the home. 'By the time I got to university and I wanted a hi-fi, I built one from scratch. If I wanted, say, a digital clock, I built one from scratch,' said Wilson, who also created an electronic cow feeder during a holiday job.[8]

Wilson generated bundles of ideas including Hawk, which swiftly became the Acorn System 1, a computer kit mounted on a pair of circuit boards aimed at engineers and lab workers. CPU's consulting work was soon eclipsed and the Acorn brand was chosen for a system designed with growth in mind – and also so the company would be listed ahead of another computer supplier, Apple, in the telephone directory. Hauser confirmed Wilson's employment after graduation by taking their parents out for a cream tea in Grantchester, a village across the river from Cambridge, and fixing their annual salary at £1,200 a year.[9]

Successor versions of the Acorn System offered improved display, memory and a floppy-disk drive, but Curry wanted the company to take a different direction. His Acorn Atom was based on the third Acorn System, and viewed as a response to the Sinclair ZX80, which Sinclair, now ensconced at Science of Cambridge (soon to become Sinclair Research), had debuted in January 1980, ready-built with a printed keyboard for just £100. Both computers rubbed along fine at different price points, both still far cheaper than the available American imports.

---

When Curry got wind of the BBC computer plan, he tried to put forward Acorn's next-generation device, the Proton, for consideration. But the broadcaster had already chosen to badge as its own the NewBrain computer, which was under development at another firm, Newbury Laboratories, and had, ironically, started life as a Sinclair Radionics project.

Like Sinclair Radionics, Newbury was state-controlled so it represented an elegant solution for the BBC, which wasn't keen to be tied to a large commercial manufacturer. However, when it became clear Newbury could not deliver on time, the contract was put out to a select few firms – including Sinclair Research and Acorn – and Curry seized his chance to meet the brief. 'We were being absolute tarts about it,' he later confessed. 'We were doing just what the BBC wanted us to do and Clive certainly wasn't doing that.'[10]

Internally at Acorn, the Proton was viewed as a compromise machine. Incorporating two microprocessors, it bridged the gap between those engineers that wanted to build a small, cheap device and those that had designs on a big, expensive workstation better targeted at the professional crowd. That gave it the range to satisfy the BBC's exacting specifications. Now all Acorn needed to do was build one – and quickly.

The wily Hauser played Wilson and Furber off against one another, calling each on a Sunday evening in February 1981 to report that the other thought it possible a Proton prototype could be put together for the BBC to view the following Friday. They protested, shrugged, and so began a marathon five days in the office, hand-drawing a scheme, sourcing components and wire-wrapping them together through the night with colleagues. 'We worked together very productively and not always smoothly but I would say creative tensions are probably a good thing,' Furber said of his partnership with Wilson, who had a pin-sharp memory but could be forthright.

At breakfast time on Friday morning, with the BBC delegation due at 10am, the machine sparked into life – but only after Hauser suggested an erroneous cable was disconnected. Wilson began adapting Acorn's operating system to create a Basic computer language just for the BBC and the fortunes of the company – still not three years old – were transformed.

There were pivotal moments to come in the creation and development of ARM. But unless Curry had pitched perfectly and unless Hauser had convinced Wilson and Furber to work relentlessly that week, there would have been no Acorn-supplied BBC Micro. And without the success the BBC Micro bestowed on Acorn, today there would be no ARM.

## The Dawn of the Personal Computer

Personal computers had long been dreamed about. On 3 November 1962, in an article headlined 'Pocket Computer May Replace Shopping List', the *New York Times* reported on John Mauchly's crystal-ball-gazing at a recent meeting held by the Institute of Industrial Engineers. 'There is no reason to suppose the average boy or girl cannot be master of a personal computer,' said the co-designer of the Electronic Numerical Integrator and

Computer (ENIAC), which had been the first programmable, electronic, general-purpose computer when it was created in 1945, initially to calculate artillery settings for the US military.[11]

What these machines could eventually do and how small they would become was a source of great fascination. Filling the basement of the Moore School of Electrical Engineering at the University of Pennsylvania, the ENIAC was far from pocket-sized.

Shrinking a mainframe computer onto a desktop was a task that preoccupied engineers during the 1970s. What today is recognised as conventional – a one-person terminal featuring graphics, on-screen symbols representing files and programs and a navigational device to move between them that became known as a mouse – first emerged from Xerox Corporation's Palo Alto Research Center (PARC) in the middle of the decade.

But what powered the flurry of product launches over the next five years or so was not the design aesthetic but the power and the price of the microchip inside them. They might have tried to differentiate themselves on retailers' shelves and in glossy magazine advertisements, but most of the home computers on the market at this time relied on the same brain.

Acorns, Apples, Ataris and Commodores, as well as Nintendo's original gaming system, the BBC Micro, and even the fruit-machine controller that was the first task for Hermann Hauser and Chris Curry's consultancy – they all contained a version of the 6502 microchip created by an American firm, MOS Technology. And in the great tradition of the industry, its launch had resulted from disaffection, and a breakaway.

Ted Hoff and Federico Faggin's original microprocessor, the general-purpose 4004, had broken the mould, finding a home in pinball machines, traffic-light controllers and bank-teller terminals. But it was Intel's 8-bit 8008 that became the first chip to be used commercially in a personal computer, not merely a calculator, when it was launched in April 1972. However, Intel was more focused on selling memory chips rather than microprocessors, so the competition had a chance to step up.

Ever since it had been founded by the Galvin brothers in Chicago in 1928, what became known as Motorola had been at the forefront of electronics, developing car radios, military communications and supplying NASA's Apollo mission to the moon. Its 6800 chip, brought out in 1974, found uses in cash registers and arcade games.

Chuck Peddle was one of the lead engineers on Motorola's 6800 and, with his wire-frame glasses and white receding hairline, looked like an expert too. He was often put in front of large industrial clients such as Ford to explain the chip's capability. They were impressed by everything he had to say except when it came to the $300 ticket price, even after applying a typical bulk discount.

'Every time we'd get through the meeting, and somebody in the room – smart guys, remember, I've got the smartest guys in the company in this room – they would say, $250 for that controller isn't going to work,' he said.[12] Customers challenged Peddle to come up with something cheaper, perhaps as low as $25 per chip. He resolved to try.

Keen to preserve its pricing power, Motorola demanded that Peddle gave up work on his low-cost alternative. Frustrated by their unwillingness to make obvious improvements – and sensing a gap in the market – Peddle jumped ship with seven colleagues, almost half of the 6800 design team. Just as Motorola was publicly unveiling the product, the defection was awkward but, for this industry, familiar.

The gang joined MOS Technology, based in Valley Forge, a small town north-west of Philadelphia. From its fabrication plant in nearby Norristown, Pennsylvania, MOS had made chips for Texas

Instruments calculators and Atari's table tennis arcade game, Pong, but needed a new product as the calculator market contracted. Valley Forge was famously where George Washington's troops mounted a fightback against the British in the American War of Independence. Peddle judged it was ripe for another revolution.

He wanted to create a chip cheap and versatile enough to be used widely, from consumer electronics to industrial machines. 'It's supposed to have been in every cash register, it's supposed to have been in every intelligent thing on the airplane, it was supposed to be everywhere,' he said later.[13] Peddle also confessed: 'It was never intended to be a computer device. Never in a million years.'[14]

The team's first attempt, the 6501 chip, was never sold for legal reasons, but the 6502 passed muster and took off. Peddle's team introduced a concept called 'pipelining' that sped up the chip's data handling, as if each instruction sat on a conveyor belt.

But the biggest change came with the system they developed to fix photomasks, the stencil that carried the chip design, rather than having to make a new one every time an error crept in. At that time, up to 70 per cent of chips were judged to be faulty and were discarded. Because 70 per cent of the 6502s that rolled off the production line were usable, the price MOS was able to charge plummeted.[15] The chip's functionality was important, but the production yield was crucial.

However, this breakthrough was initially a problem. Potential customers thought the $20 chip introduced in June 1975 at the Western Electronics Show and Convention (Wescon) in San Francisco was a scam. That was until Motorola and Intel, whose next chip, the 8080, was marketed at $150, reacted by announcing deep price cuts.

Still, the conference organisers would not let MOS sell from the exhibition floor, so Peddle took a hotel suite around the corner. He drafted in his wife Shirley to greet engineers and take their money as they delved their hands into a glass jar of chips. One of the people waiting in line was a young computer designer called Steve Wozniak, whose Apple I machine was released the following year featuring the 6502.

## The Apple Seed

When Steve Jobs saw what his best friend had come up with, he was convinced he was on to something.

Jobs and Steve Wozniak had met in 1971, united by pranks and electronics even though they were four school years apart. The pair were polar opposites: Jobs, charismatic and brattish, occasionally dabbling in Eastern religions; Wozniak, the shy son of a rocket scientist at the aerospace manufacturer Lockheed.

Both had dropped out of college. Jobs worked as a technician at Atari and Wozniak designed calculators at Hewlett-Packard. Both attended the Homebrew Computer Club, where enthusiasts shared ideas about the latest electronics.

At the first Homebrew meeting, in a garage in Menlo Park, California, Wozniak enthused over the new Altair 8800 computer and took home a sheet listing the technical specifications for the 8008 microprocessor, inspired to build his own free-standing computer. Altair used Intel's 8080 chip, but it 'cost almost more than my monthly rent' and it was hard for an individual to buy in small quantities.[16] Motorola's 6800, discounted to $40 for HP employees, was Wozniak's best option, until he heard about a new microprocessor that would be introduced at the Wescon event. It was MOS Technology's cheaper 6502 that unleashed his potential.

Jobs encouraged Wozniak to stop sharing his ideas and start selling. Early orders were assembled in Jobs' parents' garage in the Los Altos district. The name they chose to trade under, Apple Computer, was dreamed up because Jobs, an occasional fruitarian, was just back from an Oregon commune where he'd spent time among the orchards. Just as Acorn later bested Apple in the phone book, Jobs' chosen title put the new firm ahead of his employer, Atari. It sounded 'fun, spirited, and not intimidating', he said.[17]

The Apple II quickly followed, one of a 'trinity' of computers launched in 1977, the same year that *Star Wars* brought futuristic excitement to cinemas. Just like Commodore's PET and Tandy's TRS-80, the Apple II had consumers in mind, featuring colour graphics and a handful of games.

'Clear the kitchen table,' an early advert urged. 'Bring in the color TV. Plug in your new Apple II and connect any standard cassette recorder/player. Now you're ready for an evening of discovery in the new world of personal computers.'[18] Users could 'write programs to create beautiful kaleidoscopic designs' as well as 'organize, index and store data on household finances, income tax, recipes, and record collections'.

The launch was bankrolled by Mike Markkula, a small, wiry, former Intel marketer, who put vital funds into Apple and helped Jobs to develop a business plan. Arthur Rock, the legendary venture capitalist who had paired the Traitorous Eight's ideas with Sherman Fairchild's funds two decades earlier, invested too. He was intrigued by Markkula's enthusiasm after Wozniak presented to Intel's board on the merits of the personal computer. Seeing potential through the hippy exterior, Rock roped in Henry Singleton, chairman of the US electronics conglomerate Teledyne, to become a third backer.[19]

Their faith was rewarded as Apple II unit sales leapt from 2,500 in 1977 to 210,000 in 1981, aided by the popularity among business owners of VisiCalc, a spreadsheet software package that ensured the personal computer broke through as a new, mainstream product category in the US.[20]

---

The success left Jobs with money to play with but a headache. To keep up momentum into the 1980s, Apple needed to come up with something new – and preferably something more closely associated with its leader than his partner, Wozniak, who was content to operate out of the limelight.

Jobs' colleagues suggested he took an interest in what was in development at Xerox's Palo Alto Research Center. It turned out to be a hotbed of innovation on which its parent company, for whom manufacturing photocopiers was its bread and butter, had singularly failed to capitalise.

PARC researchers had created Alto, a prototype personal computer with a graphical interface and point-and-click mouse that could access information neatly organised in documents and folders. Without seeing the sales potential, Xerox had restricted these workstations to its own staff and some select US government partners.

Latterly, managers at the enterprise-focused firm decided the best way forward was to partner with someone who knew the consumer market best. The firm struck a deal to buy shares in Apple a year before it planned to list on the stock exchange – in a trade for PARC sharing its secrets.

There was internal wariness at Xerox about the demonstrations set up for Jobs and his team in December 1979, but Larry Tesler, one of the engineers to show off the Alto, viewed it as a meeting of minds. 'I was getting better questions from the Apple management than I ever got from the Xerox management,' he said. 'It was clear that they actually understood computers.'[21]

Stanford-educated Tesler joined PARC in 1973 and had worked on the programming language Smalltalk and the Gypsy word processor. But his most famous innovation was creating the cut, copy and paste function for moving text around in word documents. He strongly believed that personal computers should in future be accessible to the millions of people that had not studied computer science. It was a view decried by many at PARC who saw little potential in low-brow, hobby machines.

The shares that Xerox bought appreciated in value impressively, but Apple got the better end of the bargain. There was ultimately little in the way of collaboration between the pair as discussions to buy or license PARC's technology went nowhere.

Jobs extracted two things, however. On returning to his headquarters, staff were ordered to get working on their own graphical interface. And Tesler's arrival at Apple in July 1980, with colleagues in tow, meant his intuitive and user-friendly approach would be brought to bear on future computer models. Thousands of miles away in Cambridge, UK, a group of engineers would come to regard Tesler as a powerful ally.

## Boom and Bust

It was the closest most computer engineers would come to attaining rock-star status. Steve Furber – himself a musician in his spare time – knew the BBC Micro was going to be a huge hit when he arrived at the Institution of Electrical Engineers in central London to give a talk about it with colleagues Roger Wilson and Chris Turner.

The main lecture theatre at the Savoy Place venue, close to Waterloo Bridge on the River Thames, seated several hundred but that day in 1982 three times the capacity had turned up – some from as far away as Birmingham. 'We were booked to give the seminar two more times (and many other times around the UK and Ireland) just to meet demand,' said Furber, who had only joined Acorn formally in October 1981.[22]

What began as an accompaniment device for a TV programme quickly became a cultural phenomenon. At the start of the project, the BBC thought it might sell 12,000 machines. But, after contracting Acorn as its supplier, orders for the two Micro models, priced at £235 and £335, had already reached that level by December 1981, even before the show aired.[23]

To stoke interest, there was widespread media coverage that portrayed home computers as a passport to prosperity. But there were supply problems too. The BBC TV series had already been pushed back to January 1982 because, ironically, Acorn needed more time to produce its machines after stepping in to replace the much-delayed NewBrain.

A complication arose when microchips supplied by British engineer Ferranti to control the screen display, so-called uncommitted logic arrays (ULAs), failed on heating up. It took Acorn a frantic six weeks to work through the problem, which meant that few viewers had taken delivery of their machine when *The Computer Programme* began. The BBC also had to sanction a price increase as the cost of some components sourced from overseas soared.

The state support didn't stop there. The BBC Micro had been chosen by the Department of Industry as one of two devices that would qualify for a 50 per cent government subsidy, with the aim of installing a microcomputer in every secondary school by the end of 1982. That year had been designated as the UK's 'Year of Information Technology' and the minister responsible, Kenneth Baker, eagerly attended the computer's launch. Meanwhile, teachers dashed to incorporate the Micro into their timetables and parents followed suit by buying their children what they had seen on TV – and what many were starting to experience in the classroom.

'The computer is each pupil's own personal teacher – a teacher with infinite patience which can work at his own pace,' said the prime minister, Margaret Thatcher, in a speech on 6 April 1981 to announce the Micros in Schools scheme. 'For children all this is fun – but it also has true educational value.'[24]

Acorn produced 1,000 machines in January 1982, 2,500 in February and 5,000 in March. In October, when Acorn had just about caught up with the backlog, the government announced its discount scheme would be extended to primary schools. Some 67,000 devices had been delivered by Christmas that year.[25] The Department of Industry badgered Chris Curry to keep up.

Clive Sinclair was also dogged by supply issues for the ZX Spectrum computer, his powerful retort for not winning the BBC contract. Priced at £175 or £125, the device with a rubber keyboard still outsold the Micro heavily. Its cut-price success prompted Acorn to launch the Acorn Electron, a budget version of the Micro, in August 1983.

The following month, Acorn was steered onto the stock market with the help of the blue-blooded stockbroker Cazenove. Its flotation on the Unlisted Securities Market was small beer compared to many of the Thatcher government privatisations that 'Caz' had a hand in. But Acorn's £135m valuation – amounting to paper fortunes of £64m and £51m for Hermann Hauser and Curry respectively – marked an amazing rise for a firm whose profit of £8.6m had mushroomed from just £3,000 in 1979. The pair had put in £50 each to start CPU, and the same again into Acorn.

The company's new investors were banking on further explosive growth, but to achieve that would bring far greater complexity, in part because of the industry's global supply chain that was already in place. Memory chips had to be ordered up to a year ahead so that production facilities in Wales and Hong Kong could receive components in time for assembly. Acorn's principal route to market, mail order, gave way to stocking the high-street retailers, which meant expanding the sales team. For Christmas 1983, the Electron failed to meet demand. Curry was determined not to fall short the following year.

Home computers were glamorous now. Acorn sponsored the Formula Three driver David Hunt. He was not a patch on his brother James, the 1976 F1 world champion and renowned playboy, but the publicity was useful. Money was thrown at first-class air tickets and fancy hotels to follow Hunt's racing schedule around the world.

The company had also moved into larger premises, an old waterworks on Fulbourn Road, in the district of Cherry Hinton on Cambridge's city fringe, where a green acorn logo was mounted over the entrance. A promotional video showed the 'sheer bustle' of daily corporate life on the inside. The construction of a research and development building to the rear on which 'contractors are working to a tight deadline because Acorn needs that extra space' was earmarked to house the myriad of new projects it had embarked upon.[26]

It was ambitious – perhaps too ambitious. Acorn was forced to draw a line under a plan to expand into the US, having accumulated £6m of losses. Worse was to come now when the Micros in Schools programme ended. Apple, the company that had caused Acorn so much trouble in the US – and to which Acorn was often likened – set its sights on carving out a big share of the UK education market.

By Christmas 1984, prices were plunging and stock was not shifting. Thanks to the BBC's project, the UK had the highest penetration of home computers in the world. But now they were viewed predominantly as games machines, many consumers had moved on. When Sir Clive and Curry clashed in the Baron of Beef, compact-disc players were fast becoming the must-have electronic toy to unwrap on Christmas morning.

---

Long before shares in Acorn Computer were suspended on 6 February 1985, the City of London and financial media smelled blood. Under the headline 'Acorn's Star Fades', on 13 January the *Sunday Times* reported concerns about its future as retailers slashed prices to clear unwanted Micros and Electrons at a loss. Even the company's investment banker, Lazard Brothers, was quoted as admitting that 'sales of both machines are disappointing – the company is going through a bad patch'. The company claimed Christmas sales of 200,000 computers. Another 250,000 machines were languishing in a warehouse three miles away across Cambridge.

According to a 10 February report in the *Sunday Times*, Acorn was locked in meetings with its new financial advisers, Close Brothers, to try to secure a financial rescue. Lazard Brothers had been sacked and Cazenove, which only 17 months earlier had ushered Acorn onto the stock market, had submitted its own 'resignation in sympathy', the report said. Hauser and Curry were cast as 'stubborn individuals with a tough negotiating style', whose 'best asset was the ability to negotiate a contract'. [27]

In a deal struck in the early hours of 20 February, Olivetti, the Italian computer maker, agreed to pay £12m for a 49 per cent stake in Acorn, providing ready cash to stave off creditors. Curry, the marketing brain behind Acorn's early successes, had also diversified into publishing and quickly departed to found General Information Systems, a maker of smartcards and their readers.

Incidentally, his former employer and sparring partner, Sir Clive, had that January launched another innovation on the world: the Sinclair C5. His electric car, travelling at speeds of 15 miles per hour, was quickly ridiculed and the vehicles business was put into receivership in October, having lost him £7m. The following April, Sinclair's computing operation was substantially bought by a rival, Amstrad, for just £5m. The pub fracas 'was nothing too vigorous', Sir Clive later said. 'I think he [Curry] was pretty shame-faced but we soon patched things up.'[28]

Hauser stuck with Olivetti to lead its network of research laboratories. During the speedy due-diligence process, he had mentioned a new project his engineers were working on that was showing great promise. But it wasn't really until several months later that the new Italian leadership fully understood everything they had bought into.

## No Money, No People

A year earlier, during the spring months of 1984, key members of Acorn's engineering staff received a letter from their colleague John Horton, the technical director. The company had a reputation for being leaky, with insiders often spilling secrets to computer rivals, but now it was working on something it needed to keep firmly under wraps: Project A.

'You should be aware that if this information or even the existence of "Project A" were to become known outside the Company, it would be extremely damaging to the Company,' Horton wrote sternly. 'I am therefore seeking your written agreement and acknowledgement that you will not discuss the existence or content of "Project A" with any person outside the Company or with any person within the Company who is not part of the "Project A Team".'[29]

While Acorn was engaged in a valiant dogfight with Sinclair Research, it was already thinking of its future. The company faced a slew of competitors and, if it was to survive, it needed a microprocessor that could deliver better performance. It needed to trade up from the 8-bit 6502 chip on which much of its success had been built.

A new Acorn recruit, Tudor Brown, who had completed a master's in electrical science at Cambridge University, was handed the task of evaluating the 16-bit processors on the market from Motorola, Intel and National Semiconductor. Nothing blew him away but one option that emerged was to license and modify Intel's 16-bit 80286 chip. It had been brought out in 1982 to supply the fourth version of IBM's PC that was colonising the desktop market, particularly the business segment. In a fateful decision, Intel declined to help.

---

As they debated what to do, Roger Wilson and Steve Furber were drawn to a bundle of research papers dropped onto their desks by Andy Hopper, a well-networked Cambridge University lecturer in computer technology who was also an Acorn director.

From the University of California, Berkeley, Hopper had obtained details of a 1981 project to produce a high-performance central processing unit on a single chip: a reduced instruction-set computer (RISC). Leading the students was David Patterson, Berkeley's professor of computer science, who, a year earlier, had written a research paper, 'The Case for the Reduced Instruction Set Computer', together with Bell Labs' David Ditzel.

Microprocessors were barely a decade old, but their increasing complexity was already a cause for concern among some engineers. Instruction was overlaid on instruction, but nothing was ever stripped away. It meant the latest addition to a chip family was compatible with its elderly relatives but risked becoming slow and unwieldy.

At the same time, the microchip already inspired religious zeal. Developers would debate the versatility of the 6502 versus the Z80, for example, the two chips powering the Acorn and Sinclair computers respectively. But the pros and cons of RISC set against a complex instruction set computer (CISC) took matters to a whole new level.

A RISC prototype had been created by an IBM researcher, John Cocke, when his team were designing a microcontroller for a telephone exchange in 1975. The 801 chip, named after the building number of IBM's research unit, was rumoured to be 10 times faster than anything else available, which Ditzel thought had motivated others to experiment. 'If you didn't get into RISC now, IBM would totally dominate the industry,' he said.[30]

However, RISC's true inventor could well have been Seymour Cray, the Wisconsin-born 'father of the supercomputer', whose models used similar architecture to Cocke without taking the RISC name. At Stanford University, Patterson's opposite number, John Hennessy, was leading RISC work too, with the snappily titled Microprocessor without Interlocked Pipeline Stages (MIPS) project.

RISC advocates argued that 80 per cent of the time the chip typically uses just 20 per cent of its instructions, so it was better to take out the rest to focus efforts on the ones being used most often. Because they broke everything down into a handful of simple instructions, RISC chips were lightning-fast and consumed less power compared to hungry CISC chips, which took time to sort through complex tasks.

Patterson and Ditzel had been investigating 32-bit architectures, which represented another leap in computing power. In their paper they challenged the prevailing idea that the increasing complexity of computers had a positive impact on the cost-effectiveness of newer models, suggesting that complex instruction sets were often used as evidence of a better computer in marketing materials. 'In order to keep their jobs, architects must keep selling new and better designs to their internal management,' they wrote. 'The number of instructions and their "power" is often used to promote an architecture, regardless of the actual use or cost-effectiveness of the complex instruction set.'

They were equally damning about 'upward compatibility', the notion that designs were mainly improved by adding new and more complex features. 'New architectures tend to have a habit of including all instructions found in the machines of successful competitors, perhaps because architects and customers have no real grasp over what defines a "good" instruction set.'

In conclusion, they added: 'We see each transistor as being precious for at least the next ten years. While the trend towards architectural complexity may be one path towards improved computers, this paper proposes another path, the Reduced Instruction Set Computer.'[31]

Not everyone agreed. 'There was a huge amount of skepticism,' Ditzel recalled of the reaction they got, but Wilson and Furber were intrigued.[32] They were also eager to see if there was still life in the 6502 that had served Acorn so well.

---

The early flurry of interest in Chuck Peddle's chip had failed to preserve the independence of MOS Technology. Even before the home computer boom that the 6502 catalysed, Commodore, which was initially one of MOS's calculator clients, bought the business in 1976 in a defensive move.

TI, which had made an early splash with pocket radios, was on a mission to regain its market-leading position in calculators. Its chips were made by MOS for customers including Commodore and it cratered an already fragile market by launching its own calculator priced at less than Commodore's manufacturing cost alone.[33]

The impact was twofold. Commodore quickly expanded into home computers, while still licensing the 6502 to allcomers, including Acorn. But with funds tight, further development of the chip was a low priority.

Bill Mensch, a layout engineer who had been hired into Peddle's team back at Motorola, quit MOS and returned home to Mesa, Arizona, on the fringes of Phoenix. It was there he set up the Western Design Center (WDC) to further the 6502's development.

When they visited in October 1983, Wilson and Furber were underwhelmed. Not only did they think the 6502 had reached the end of the road, but the WDC was no more than a handful of youngsters operating out of a suburban bungalow. 'We also came away fairly convinced that if they could build a processor, then we sure as hell could,' said Wilson.[34]

With Hermann Hauser's backing, the Acorn RISC Machine (ARM hereafter written as Arm in this book), project was born in October 1983. Codenamed Project A, its target was a lofty one. Rather than coming up with their own 16-bit chip, Wilson and Furber would attempt to leapfrog everything else in the market by designing a 32-bit processor that would deliver ten times the performance of the BBC Micro – and at a similar price.

It was a giant step. The team had designed part of the ULA for the BBC Micro but never a whole processor before. They resolved to see if they could take a model of largely academic interest and propel it into the commercial world. Its simplicity and low cost were informed by the two things Hauser gave them to work with: no money and no people. 'I always felt that my biggest contribution was to let it happen,' he said simply.

Other RISC projects were fixed on producing high-end business workstations but Wilson and Furber thought it would be interesting to attempt something simpler. They coined a catchphrase, 'MIPS for the masses', where MIPS stood for 'millions of instructions per second', with the idea that cheap computing power should be within reach of as many people as possible.

On 26 April 1985, the first Arm-based microchips arrived back from Acorn's manufacturing partner, VLSI Technology in San Jose, California, and were running BBC Basic straight away. When the computer calculated pi accurately, champagne corks were popped. Ready to lift the veil of secrecy on Project A, Furber rang a journalist with the story of his new processor only to find they didn't believe him.

The Arm1 featured 25,000 transistors, a tenth of the number carried on the latest Motorola chip, and Furber's tests showed it still outperformed. And it was simple, with a reference model written in just 808 lines of Basic. Even better, the chip appeared to work without any power, registering a zero reading on an ammeter that Furber hooked up, instead drawing juice from various adjacent components. The design was focused on keeping the cost down to the extent that it used cheap plastic packaging rather than the more expensive ceramic setting used by some other processors. Rather than 10 times the performance of the BBC Micro, they had come up with something 25 times better.

Success came at a vital time. On 13 August, trading in Acorn shares resumed after a seven-week suspension while a further refinancing package was thrashed out with Olivetti, which increased its stake in the business to 80 per cent from the initial 49 per cent.

Furber eventually found someone to believe him. Under the headline 'Acorn beats world to super-fast chip', *Acorn User*'s October 1985 edition reported that: 'The performance of the chip – which worked first time – was undoubtedly a factor which encouraged Olivetti to stay with Acorn. News of the success came when negotiations on the refinancing package were at a critical stage.'[35] .

A second version of the chip, the Arm2, found a commercial home in the Acorn Archimedes computer in 1987, marking a decent success for Acorn. However, it wouldn't be long before Arm broke free from the desktop altogether.

## Navigating the Future

The president of Pepsi, John Sculley, had not succumbed to several attempts by Apple to recruit him, but Steve Jobs' immortal line proved irresistible. 'Do you want to spend the rest of your life selling sugared water or do you want a chance to change the world?' Jobs asked Sculley, a high-flying corporate leader, 16 years his senior, as the pair looked out from a penthouse balcony high above Manhattan one day in March 1983.

Contributing leadership experience, professional discipline and marketing nous, Sculley started work as Apple's chief executive the following month, with Jobs as chairman. But the pair spent little more than two years together before the board sided with Sculley, who wanted to remove Jobs from the helm of the underperforming Macintosh division.

The Macintosh computer was Jobs' pride and joy, launched in a blaze of glory with an expensive, Orwellian advertising campaign that first aired during the 1984 Super Bowl. Through its development it had quickly overtaken in his affections the Lisa, a desktop PC launched the previous year inspired by what Apple gleaned from its fact-finding trip to Xerox. Yet after the early splash, Mac sales were slow in a market that was being dominated by IBM PCs or their derivatives. During the dire Christmas 1984 season that severely damaged Acorn Computer and Sinclair Research in the UK, it was actually the Apple II that powered the company through. The older device retained a strong following internally, causing a rift with those that believed the Mac was the future.

Sculley and Jobs professed to have built a great friendship but their combination was ultimately a messy culture clash. Sculley, whose claim to fame was presiding over the creation of the Pepsi

Challenge, a blind cola taste test that was the cornerstone of the brand's advertising efforts, was polite and considerate and ceded too much power to Jobs, who inspired fierce loyalty despite some everyday rudeness. 'He didn't learn things very quickly,' Jobs said of Sculley with his typical bluntness, who he decried for not embracing Apple's products. 'And the people he wanted to promote were usually bozos.'[36]

———————

After Jobs' resignation in September 1985, Sculley steered a recovery in Apple's sales. Looking further out, in 1986 the company created an Advanced Technology Group (ATG) to hunt for and incubate cutting-edge ideas. Jobs the visionary was gone, as was a disillusioned Steve Wozniak. ATG looked and felt a lot like 'AppleLabs', the mooted development division Jobs could have run if he had opted to stay on, having been freed from managerial responsibility. Larry Tesler, recruited six years earlier in the 'raid' on Xerox and having earned a reputation internally as courageous yet cool under pressure, was elevated to vice president. Soon he stepped in to run ATG.

Sculley, meanwhile, fell under the influence of Alan Kay, another Xerox PARC alumnus, who in 1984 had been made an Apple Fellow, part of a scheme designed to tie computing innovators to the firm. In the 1960s Kay had dreamed up but never built a device he called Dynabook, a computer primarily for children with a keyboard, touch-screen and stylus for writing, that could be carried like a notebook and link to larger computers via radio frequency. In a 1972 note, the moustachioed Kay speculated about 'the emergence of personal, portable information manipulators' powered by one or, at most, two chips that 'typically contain the equivalent of several thousand transistors'.[37]

Portable, powerful, all-pervasive: it was a familiar dream that the technology was racing to catch up with. Jobs had enthused about Dynabook in conversation with Kay, and its ambition was not dissimilar to what the ENIAC co-designer John Mauchly had spoken about. The Dynabook informed Xerox's Alto prototype and its influence emerged again, in the epilogue to Sculley's 1987 autobiography *Odyssey*, which was subtitled: 'Pepsi to Apple . . . a journey of adventure, ideas and the future'.

Sculley wrote: 'A future-generation Macintosh, which we should have early in the twenty-first century, might well be a wonderful fantasy machine called the Knowledge Navigator, a discoverer of worlds, a tool as galvanising as the printing press.' Individuals could use it 'to drive through libraries, museums, databases, or institutional archives . . . converting vast quantities of information into personalized and understandable knowledge.'[38]

How the Navigator looked didn't matter, Sculley added. 'Indeed, within the next decade, the most powerful personal computer available today will be "invisible"; like a motor; it will fit into a machine the size of a pocket calculator.' The Apple II chip 'is already small enough to wear on an earring', he observed.

The idea for Knowledge Navigator fed into a keynote speech Sculley gave later that year, and an accompanying video showed a college professor conducting various tasks on a tablet-like device. Such ideas were already being pursued within Apple, and Sculley let them flourish.

One engineer, Steve Sakoman, had worked on the HP Portable, Hewlett-Packard's pioneering but pricey laptop, before he joined Apple in 1984. To stop him leaving after working on the second-generation Macintosh, Apple's vice president for product development, Jean-Louis Gassée, allowed him to pursue his ambition of designing a pen-based mobile computing device. Sakoman called it Newton, after Sir Isaac Newton, the famous

British mathematician who featured in Apple's original logo for his formulation of gravitational theory inspired by falling fruit.

After a disagreement with Sculley prompted the departure of both Gassée and Sakoman, in early 1990 Tesler began leading the Newton project, 'to see if I could save it and turn it into something a little more practical than what they had been doing'.[39]

Newton had got out of hand. After three years in development, what Tesler found was too big, too expensive, and impractical. 'I urged the Newton team to bring the target price down from $6000+ to $4000, then to $2000 or less,' he said.[40]

Part of the reason the device was so costly was that it was running on three microprocessors. Aware that no chip existed capable of powering the Newton, Sakoman had gone to AT&T, where David Ditzel, David Patterson's RISC collaborator and former PhD student, was experimenting with its C-language Reduced Instruction Set Processor (CRISP) design.

AT&T had a large chipmaking division, which until the break-up of the US's monopoly telephone system in 1984 had only produced for internal use. Now it was adjusting to supplying the outside world.

Apple had agreed to fund the development of what became known as Hobbit, a low-powered version of CRISP, in exchange for a period of exclusive use. It had poured an estimated $5m into the project, but Tesler's view was that the Hobbit was 'rife with bugs, ill-suited for our purposes, and overpriced'.[41] Just one Hobbit was nowhere near powerful enough to run all of Newton's software. AT&T demanded several million dollars more to fix it, so Tesler began exploring alternatives.

———————

Arm was already in the building. A software engineer, Paul Gavarini, joined Apple's ATG in November 1985 from Zilog, the chip company whose Z80 had powered Sir Clive Sinclair's computers. On a visit to Olivetti in Milan, he heard about an interesting chip design being worked on in Cambridge, at Acorn Computer, by now Olivetti's subsidiary. With carte blanche to create a new machine for Apple, Gavarini returned from Cambridge in early 1986 with a couple of Arm chips in his pocket.

RISC was still not to everyone's taste. The year before, Gavarini's colleague Tom Pittard had hosted a presentation by a team from MIPS Computer Systems, the company born from John Hennessy's Stanford project, but no one at Apple latched on to their work.

Arm was more fortunate. After some experimentation, Gavarini and Pittard created a device that ran both Apple II and Macintosh code – effectively uniting the two sides of Apple – as well as Microsoft Windows. The so-called Möbius Project was clever, but it was too much for the company's warring factions, for whom the street between their two buildings on Apple's campus had been christened 'the DMZ', short for demilitarised zone.

Matters came to a head when Gavarini presented his machine alongside two others for Apple management, including Sculley, to evaluate. The Möbius was running four times faster than the second version of the Macintosh, Gavarini recalled. Nevertheless, he said, 'they all preferred Motorola chips', which Apple had adopted for the Lisa and Macintosh series. About switching to another chip: 'They didn't want to hear it.'

Möbius was canned, and Gavarini said he received a letter co-signed by Gassée demanding that he stop working with the Arm chip. The project was boxed up and lay mouldering on the floor of Pittard's office.

Sometime later, Tesler called, asking after the chip's performance data. At a briefing to go through their findings, Pittard slid an Arm chip across the table. Now Tesler was interested.

## A Conversation in the Desert

After several years spent advising engineering firms on how to introduce computers to their design and manufacturing work, Malcolm Bird liked the idea of switching inside a business so he could put his own ideas into practice.

But in February 1990, three months on from joining Acorn Computer as its technical director, he was just about out of options for one particular task. The former manager at PA Consulting, a well-known firm originally founded to help the UK war effort by boosting worker productivity, had a broad array of projects in mind on his arrival. One early piece of work was to develop the products to take advantage of Arm another, of mounting significance, was to find someone willing to fund it further.

This last quest had seen Bird travel all over the world. He was working through a list of contacts provided by Elserino Piol, the Olivetti executive and Acorn chairman who regularly darted into the Cherry Hinton office on business trips from Italy, demanding progress updates through plumes of cigar smoke. Acorn, and Olivetti, sensed Arm's promise but did not have the money to spend to really capitalise on it.

In the race to dominate the market for RISC processors, the smart money was on MIPS Computer Systems, founded by John Hennessy's Stanford team. The company was on the up. After selling designs to US manufacturers Digital Equipment Corporation and Silicon Graphics among others, it had listed its shares in December 1989 and embarked on the more ambitious plan to become a computer vendor by making devices based on its own chips. At Sun Microsystems, where David Patterson worked as a consultant, the SPARC (Scalable Processor Achitecture) instruction set had also made good progress with the original Berkeley RISC design.

The Acorn Archimedes, the original reason for Arm's creation, had been well received since its launch in June 1987, but the technology needed to push on if it wasn't going to run out of road like the 6502 chip had done.

The industry was curious about Arm but that did not extend to backing a technology buried in a mid-sized Cambridge company whose best years appeared to be behind it. In particular, Bird was surprised that the European chipmakers, the French-Italian SGS-Thomson and Philips of the Netherlands, did not display the vision or confidence to become involved in something grown close to home.

Acorn knew there had been some interest in Arm from Apple, chiefly because the American company was buying development systems from its manufacturing partner and licensee, VLSI, to evaluate the processor. These purchases were followed by deep technical questions shrouded in great secrecy. By the time Acorn understood that Apple was working on a device that needed a high-performance, low-power chip, Bird struggled to get a meeting and the assumption was the project had been dropped or Apple had found an alternative.

Now, as he sat in Acorn's deserted office on Newmarket Road late one night, Bird's thoughts turned to the inevitable: if Arm could not be sold off or a funding partner found, how best to wind it down, while keeping an eye out for rescue opportunities. 'It's pretty clear we wouldn't have been able to continue funding development on our own,' Bird said. 'If anything, development needed ramping up in order to keep up with what was going on elsewhere.'

And then the phone rang, and everything changed.

———

Bird picked up the receiver to Hermann Hauser, Acorn's co-founder, who was calling from the US where he had a lead.

Ever the networker, Hauser had bumped into someone useful from Apple at a technology conference.

Esther Dyson's gatherings were must-attend events for the US computing industry. The elfin editor of the influential industry newsletter *Release 1.0* convened the annual Personal Computing Forum, which had rapidly become the elite high-tech conference where innovations were spotted and deals were made. 'She sees in the computer a product which brings together all of the exciting intellectual possibilities that can shape the thinking of the world,' wrote John Sculley.[42]

That January, more than 400 luminaries and hangers-on crowded into the flashy Westin La Paloma resort in the desert hills outside Tucson, Arizona, to discuss the future and talk up their own projects. Among them, Steve Jobs bemoaned the pedestrian sales of his $9,995 NeXT computer, his post-Apple project, when compared to the brisk launch months of the Macintosh. And on the sidelines, Hauser buttonholed Larry Tesler.

Hauser was still a director at Acorn and had done his best to shield Arm from spending cuts, but, after leaving his Olivetti research role, spent most of his time on a new start-up, the Active Book Company (ABC), into which he had sunk £1m of his own money. ABC had created a prototype for a book-style computer that made use of the Arm design, which Hauser, dipping into his expansive contacts book, was showing anyone willing to look.

Tesler had followed Arm's progress at a distance since reviewing the chip's performance with Tom Pittard. But a deeper relationship was frustrated in part because Arm was locked inside one of Apple's computing competitors. Still, he hadn't identified an alternative to the disappointing Hobbit chip for the Newton. At the PC Forum, Hauser thought it was worth another try. 'Is it too late for you to adopt Arm?' he asked Tesler. 'What do we need to do to make it happen?'

Sensing an opportunity, Hauser briefed Bird on the conversation and passed on Tesler's contact details. Bird followed up immediately. When they spoke, he reassured Tesler over his greatest concern: Acorn was willing to separate out Arm into a new, independent company.

Tesler liked what he heard. So began a period of intense activity.

## Chapter 4

# THIRTEEN MEN EMBEDDED IN A BARN

## A Place in the Country

The quiet village of Swaffham Bulbeck in East Cambridgeshire put on its first Gilbert and Sullivan production in 1982. Locals chose to perform the one-act comic opera *Trial by Jury*, an early product of the immensely successful collaboration between the Victorian dramatist and composer.[1]

The show was staged in the Long Barn in June, which was a quiet time for rearing turkeys, the building's primary purpose. At its peak, David Rayner's farm was raising 25,000 birds a year and turkey plucking was a handy way for young villagers to pick up some Christmas spending money.[2]

Rayner and his wife Brenda had become popular figures in the community since moving to Swaffham Bulbeck with their young family in 1962. The village's long history was enshrined in its name: Swaffham recalling the Swabians, a tribe that migrated to the region from south-west Germany in the sixth century; Bulbeck suggesting it was once held by the French Bolebec family whose members arrived after the Norman Conquest and founded a convent locally around 1150.

The Rayners bought and renovated the dilapidated Burgh Hall, a Grade II listed, 15th-century manor house surrounded by a moat on the southern edge of the village. Like most farmers, Rayner had to be entrepreneurial to survive. He soon gave up on cattle and sheep on his adjacent 2,500 acres because of poor financial returns and poured his efforts into developing the Scotsdales garden centre he had acquired to the south of Cambridge.

In time, new regulations prevented him raising livestock in the way he liked, so Rayner's gently undulating farmland became purely arable: wheat, barley, rapeseed and beans. The switch meant he had less use for the cluster of four old farm buildings, including the Long Barn and, at the back of the yard, Harvey's Barn. Modern machinery was better housed in easy-to-access sheds.

By the late 1980s, Rayner had resolved to turn the barns into office buildings to cultivate a new revenue stream. Out went the thespians whose performances had become a staple of the village calendar, out went the poultry and sacks of provisions. Harvey's Barn – the origins of whose name were lost in the mists of time – was simply converted but had its old timber frame preserved as part of securing planning permission. One potential tenant had agreed terms but the deal fell apart so that in early 1991 the property lay vacant.

Every company starts somewhere. It is the moment an idea becomes a product that spawns a strategy, made real with investors' money and staff to house. Apple's founders began assembling computers in Steve Jobs' parents' garage in Los Altos, California. They may have got the idea from David Packard who, a generation earlier, welcomed Bill Hewlett to his garage in Palo Alto and another chapter of Silicon Valley folklore was written.

Looking out over the fens, Harvey's Barn might have been a short drive from Cambridge but for visitors to Swaffham Bulbeck spying the thatched and timber-framed houses it was like step-ping back in time. Still, the 13 men that set up shop here were more worried about forging a future than dwelling on the past. The barn was cheap, available and it seemed like a particularly English solution to their need for premises. In fact, its quaintness would serve them well.

And so it was in March 1991 that a new name was added to the list of Swaffham Bulbeck's residents: Advanced RISC Machines Ltd.

## A New Backer and a Business Plan

Just because Malcolm Bird had heard encouraging noises from Apple in late February 1990, it didn't mean Arm's future was guaranteed. The advanced research and development (R&D) team that housed Arm within Acorn knew that their bosses and Olivetti executives had been furiously trying to find an investor or buyer for a while, so Bird did not want to raise expectations. However, he knew he had to act fast if he was to convert Larry Tesler's interest in the technology, because key staff were getting itchy feet.

It was almost five years since the Arm chip design had first been produced – a lifetime in the semiconductor world – and the team looked at the impressive progress of RISC processor rival MIPS Computer Systems with envy. 'Several of us were getting fed up,' said Tudor Brown. 'We had created this great technology but it was going nowhere. The lack of direction was very frustrating.'

The members of the advanced R&D team were a close-knit bunch, so if any of them left they knew it would let the others down. By this point, that didn't stop them looking elsewhere for work out of desperation. The disintegration of the Arm project was a real possibility. 'The technology would have just gone into a cupboard and we would all have gone our own ways,' said Mike Muller, another member of the team.[3]

Inmos, a promising British microchip firm, had been sold to the French-Italian group SGS-Thomson the previous April. Hopes had been high for the venture – and the wider UK industry it might catalyse – when it won £50m of support from the UK government's National Enterprise Board in 1978.

Inmos's hero product was a 'transputer', a 32-bit device that overcame a bottleneck in the fabrication of better-performing central processing units (CPUs) by supporting 'parallel' processing. That meant the chip could handle more than one instruction at once – as well as linking with other processors to boost capacity.

Conceived in Bristol and manufactured at a plant in Newport, South Wales, that had been designed by the notable architect Richard Rogers, it was hailed as a British success story and found a home in early video telephones, laser printers and NASA's ground control system. However, Inmos had to battle the suspicion its products were more suited to scientific applications than high-volume consumer use.

After it passed into the hands of the conglomerate Thorn EMI, investment demands rose inexorably. From there Inmos followed the familiar path trodden by UK technology firms into foreign ownership. In fact, the sale teetered on embarrassing, according to one hard-to-verify story. According to an urban myth, SGS's charismatic Italian chief executive Pasquale Pistorio stood before his new workforce at the Aztec West site in Bristol on the day of the takeover and, with a slip of the tongue, said how pleased he was to be acquiring Intel.

Amid much hand-wringing over why the UK could not sustain its own microchip industry, at least Inmos now had the funds to launch an improved version of the transputer. On steadier ground in the press statement, Pistorio declared: 'We intend to put all the financial and marketing muscle of SGS-Thomson in support of Inmos to establish the transputer as a world standard.'[4]

Part of his plan was to develop a digital signal processor (DSP) that translated sound and images from analogue and would become a vital component for mobile communications. To do so, SGS was preparing to pick off several members of the Arm team who were craving resources.

A sense that the game was up was compounded in April 1990 when Steve Furber, the co-creator of the Arm architecture, announced he was departing to become the professor of computer engineering at the University of Manchester. Given he was only 37, the job application had been a long shot, but Furber was vexed that Arm's development was falling behind the competition and was eager to try something new close to where he had been brought up.

When he departed, he had no knowledge of Apple's interest, having abandoned his own efforts to construct a business plan for the chip design. 'You would need to sell millions of the things to cover the company's business costs and that was way beyond our plausible expectations,' he said.

---

It was too late to stop Furber, but Bird somehow convinced everyone else to stay put, at the same time as making all the running with Apple. From May 1990 onwards, with the vigorous backing of the Acorn board, he travelled regularly to the company's headquarters in Cupertino, California, to thrash out an agreement.

On one occasion, a short man ambled into the room, perched up on a counter and listened in to discussions for a few minutes without saying a word. Bird realised later it was Apple's chief executive John Sculley. The company remained vague about what specifically it wanted to use the Arm design for, but the project clearly reached right to the top.

At the same time as having a corporate discussion, Apple was assessing the credibility of Arm's engineers. 'They wanted to know we understood what their problem was, what they wanted to change and for them to believe that we knew how to do it,' said Muller.[5]

Apple could afford to buy Arm outright but the idea of a joint venture with Acorn quickly solidified because neither party wanted the other to have more control. Tesler preferred to keep Apple's interest in what would become a new company at below half because that meant the reporting requirements were simpler. He agreed to put in £1.5m for a 43 per cent stake.

Acorn took the same, although its contribution came from the value of the intellectual property and the people it injected into the venture. At a cost of £250,000 plus use of its design tools, VLSI took a small minority. Apple insisted that the 'Acorn' in Acorn RISC Machine was dropped in favour of 'Advanced' because it didn't want to be seen dealing too closely with a perceived rival.

When the investment proposal went up to Apple's board, Bird was called over again to convince them individually that backing a UK start-up was a good idea. Around the table were leading industry names who had carved out reputations as canny investors. The board still included the trio that gave Apple its start in the 1970s. The company's chairman was Mike Markkula, the former Intel marketer, who for a time also served as Apple's chief executive. Alongside him sat the venture capitalist Arthur Rock and Teledyne's Henry Singleton.

For such a modest sum, their assent was never in doubt, but the final agreement still came down to the wire. Tesler cannily kept two processor options alive just in case the spin-off foundered. At the eleventh hour, VLSI tried to negotiate up its shareholding but Tesler was having none of it. During a crisis call in early November, he told VLSI's vice president Cliff Roe to back down or Apple was pulling out, and then slammed down the phone.

'Do you really want to kill this whole deal?' Bird said, as the pair were left hanging. Roe quickly recanted and Bird called back Tesler, who returned to his meeting to confirm to colleagues that the top-secret Newton device would be using a processor based on Arm designs.

It was a moment to savour. From a broad field of RISC developers, Arm had been preferred ahead of AT&T, one of the largest chip firms in the world. To rub salt in the wound, since late 1983 AT&T had owned a 25 per cent stake in Olivetti, Arm's parent company, for what good it had done them in trying to hold on to the business itself.

'I was fascinated by the Arm for no good reason,' Tesler said. 'It just had a very unique and elegant instruction set, that was kind of RISC, but kind of not RISC at the same time.'[6]

Arm was extremely convenient. Apple did not have a lot of choices because it was not a volume player and was experimenting in an unproven category. Tesler added: 'We wanted a custom version from whatever company we worked with. And we couldn't get the big companies working with us, because they didn't want a custom version for Apple. We're the smallest market.'[7]

And Arm was cheap. The £1.5m stake to set it up as an independent company, equivalent to $2.5m, was 'less than what we would have paid to AT&T', Tesler admitted.[8]

Sometime after he settled into academia in Manchester, Steve Furber travelled to New Jersey to give a seminar for AT&T. He recalled: 'They had an Apple poster on the wall that they were using as a dartboard.'

Other than perhaps VLSI, everyone had what they wanted: Apple an alternative to AT&T, Acorn an ongoing stake in a promising technology it couldn't afford to fund itself, and Arm a shot at independence. They couldn't know it at the time, but it also marked the start of Acorn and Arm's sharply diverging fortunes.

But the mood was not exactly euphoric. 'People think it was a glorious entrepreneurial start but we felt quite unloved,' said John Biggs, one of those chosen to join Arm, who began working at Acorn during university vacations and had co-authored the Acorn Electron user manual. 'Acorn was tightening its belt and chip design was a luxury they couldn't afford.'

---

The soft-spoken and dependable Jamie Urquhart was made general manager of the new venture. He was one of three designers in the advanced R&D team who shared a single Apollo computer that could run the chip-design software supplied by VLSI. For a long time, Urquhart arrived at 5am to work on RISC circuits and handed over to a colleague at noon. What the Apollo was capable of was limited. Engineers had to travel all the way to VLSI's Munich office to carry out design checks on the chips that were produced.

Urquhart carved out Arm's intellectual property from Acorn and whittled the team that worked on it down to 12 people. They largely self-selected, especially in areas of software and systems design, and were judged to be the minimum number needed to make the venture a success.

When it was clear the spinout was happening, the new Arm team was moved to sit together at the top of the Silver Building, Acorn's R&D unit behind the Fulbourn Road headquarters. If anyone at Acorn had not noticed change was afoot, they did when the team took the windows with them from downstairs because they were specially coated to reduce outside glare. At the last minute, one person who didn't join the group was Sophie Wilson, who pledged loyalty to Acorn.

The company was incorporated on 16 October 1990 under the name Styletheme Limited – an off-the-shelf shell bought from Companies House. It was struck as a joint venture between Apple Computer (UK) Limited, Acorn Computers Limited and VLSI Technology Inc. On 22 November Styletheme changed its name to Advanced RISC Machines Holdings Limited.

The following week, on 27 November, a bullish press release announcing the company's formation made clear it had to look beyond Apple and Acorn for business. Its mission was to 'address and attack the growing market for low-cost, low-power, high-performance 32-bit RISC computer chips'. Its strategy included supplying 'personal and portable computers, telephones, and embedded control uses in consumer and automotive electronics'.

The statement noted that to date 130,000 Arm-based chips had been shipped, 'placing it among the leading processors', and its products were licensed by VLSI and Sanyo Electric of Japan. Some of the chips had gone into Acorn machines and VLSI had also designed some embedded controllers. The chip's third iteration, Arm3, could process 20 million instructions per second. Tesler said there was a 'need for a new product and standard', while Acorn's managing director Sam Wauchope, a moustachioed Scotsman, talked of plans to 'pursue an aggressive product roadmap'.

Just as the particulars of the design product they had come up with were determined by harsh economics, so was how it was sold. Acorn never planned to manufacture its own chips, and nor could it really afford to, although they were designed by hand to a great degree of specification for VLSI to make. The question was how else the designs could be adapted – and who would buy them.

The launch was big news for this particular corner of the UK technology industry, but few outside that gave it a second glance. That same day, the UK was gripped by the political theatre of John Major seizing victory in a leadership election to succeed Margaret Thatcher as prime minister.

On Arm's second day as an independent company, the woman whose government championed homegrown hardware – and had ensured that a huge volume of Acorn computers went into British schools, ultimately generating the funds that gave a little breathing space to develop a new chip design – went to Buckingham Palace to tender her resignation to the Queen.

It was a new dawn for the country, as well as for a dozen expectant engineers. Now all they needed was someone to lead them.

## Enter the Salesman

On a foggy night in mid-December 1990, a group of men trooped into the Rose & Crown, a cosy 16th-century pub in the village of Ashwell near Baldock in Hertfordshire. Reduced visibility on the roads leading south-west out of Cambridge had made it a slow, treacherous drive. As they bought drinks and settled in, their guest made sure they knew he did not like to be kept waiting.

'You're four minutes late,' he snapped. 'If it had been five minutes, I'd have gone.'

As getting-to-know-you meetings went, it was a testy start that raised eyebrows among the colleagues. After all, this wasn't a social call or another hunt for fresh investment: here was the man auditioning to lead the band of engineers that formed Arm.

Robin Saxby was impatient, enthusiastic, confident, dishevelled, brusque and something of a namedropper. He was a travelling salesman with grey-blue eyes and mousy brown hair plastered down across his forehead. And he had a knack for forming relationships built to last.

As they sipped their pints, the Arm team soon found that Saxby's bark was worse than his bite, one of the many contradictions about him. Like a teacher meeting a new class, his opening gambit was designed to assert himself, and channelled some of his worry that Cambridge technology workers were too arrogant to think much of the importance of punctuality and meeting customer needs. But the engineers needn't have worried about him storming off – Saxby was far too intrigued for that.

Arm had a path to independence and a chip design with some potential, but no leader. The task of finding a chief executive fell to Malcolm Bird and Larry Tesler, who appointed a headhunting firm, Heidrick & Struggles, to conduct a search.

Saxby soon bubbled to the top of the list. A veteran of the American firm Motorola, he knew the semiconductor market inside out and how to evangelise about something new. In this industry, customers were intelligent, but they also responded to the flair of a good salesman. Too often, products failed because clients didn't understand how they could be used and they weren't willing to experiment.

———————

Born in 1947, Saxby grew up in Chesterfield and developed an early fascination with electronics. At the age of eight, he had been given his own kit to play with and by 13 he was repairing radios and TV sets for neighbours who placed orders with his father, a factory security guard, in the pub.

While studying at Liverpool University in the 1960s, he built amplifiers for rock 'n' roll bands and organised dances in his hall of residence. On his way towards an electronics degree in 1968 Saxby's final-year essay focused on the advent of colour television. But before graduating Saxby turned down the offer to work for the BBC, reasoning that 'I like to have fun and I can have more fun working for a small, lean, mean organisation than a big one'.[9]

He started out at the consumer electronics firm Rank Bush Murphy, designing colour TV receivers that used integrated circuits featuring just 50 transistors on them, and then briefly at Pye, before the lure of a better salary and company car – a white Cortina with black upholstery – took him to Motorola just as the company was venturing into Europe.

Saxby found the company culture more go-getting than at most British firms he knew. From design he was switched to become a sales engineer, helping to commercialise products and explain their technical features to customers, and from the consumer side he went to the computer division. Saxby understood that success came from responding to customer pull, adapting the technology he was selling to suit their needs. His abilities were matched only by his vaulting ambition.

Itching to run something, and reluctant to move his family abroad to take up a more senior role, Saxby quit for Henderson, a garage-door company that needed a boss to sort out its security division. Differences of opinion with his new boss meant it was short-lived.

'I am fundamentally optimistic,' Saxby said. 'One of the things I say is every problem is an opportunity in disguise. And if you don't try something, you don't learn.'

Next came European Silicon Structures (ES2), a venture aimed to develop a nascent 'e-beam' technology that would replace the photomasks used to print designs on silicon wafers with a cheaper and simpler alternative. Having raised $100m from some of the biggest European names in the industry – Olivetti, Philips, British Aerospace, Saab, Telefonica – ES2 was viewed as a test of collaboration between national champions as the European Community prepared to tear down trade barriers in 1992. Saxby was excited by the prospect of a homegrown start-up to rival the might of Corporate America. The trouble was the technology did not work and the company was burning through cash at pace.

Saxby was toiling away, trying to fix ES2's US arm, which was being dragged down by conflicting agendas, egos and over-spending, when the call came from Heidrick & Struggles with an interesting proposition for him.

---

At first Saxby was not keen. The involvement of Olivetti reminded him too much of his current situation. But Apple made his ears prick up, and the licensing model that Acorn and Apple had in mind chimed with a business plan he had written while at Motorola to spin out the microprocessor unit based on offer-ing design services to a wide range of customers. This could be the chance to test out something similar. Better still, Arm was a blank canvas on which to put some of his management ideas into practice.

It was a small industry and Saxby knew Acorn well. In fact, if he had been a more effective salesman, there might not have been an Arm chip design – or a CEO vacancy – at all.

At Motorola, Saxby was based in Wembley, north London, but one of his sales techniques was to fly prospective customers up to Glasgow and tour them around the microchip plant at East Kilbride, to the south of the city. He was forceful and enthusias-tic, waving his arms around to make a point. But after showing off the investment that had been made at the facility and empha-sising the finer points of the product, there was always time to unwind over drinks and dinner.

On one occasion, he brought Hermann Hauser and Chris Curry to Scotland with a view to supplying the next Acorn com-puter with Motorola's 68000 chip. The pitch continued on the flight home. Curry headed to the back of the plane and Hauser, who had heard enough, pretended to fall asleep on Saxby.

They had known each other for a little while. Saxby had sold Acorn's precursor company, CPU, the chips for its initial fruit-machine project. And then, when Acorn needed help to win the all-important computer contract from the BBC, Saxby had written a supportive letter. So perhaps he had – indirectly – helped to birth Arm.

A friendship with Hauser developed, so that the pair enjoyed ski trips together. Saxby was often paired on the slopes with Hauser's future wife Pamela. The Kiwi and Brit could keep each other company as they descended at their own pace, the skilled Austrian skier joked as he whizzed by.

Given their connection, it is surprising that Saxby did not consult Hauser about the Arm opportunity. He talked to many of his friends and former colleagues about whether it would work, including John Berylson at Advent International, an ES2 investor who rooted for him with Apple.

It seemed the stars were aligning. On 5 December 1990, after a fruitful first meeting with Tesler at Acorn's Newmarket Road base, they and Bird left on their way to dinner. Saxby's car was parked up outside, and the trio noted that all three of them were Saab 9000 drivers. 'He was obviously going to fit,' Bird joked.

As well as being persuaded by the technology, Saxby needed to be persuaded by the people. After his Henderson experience, it was important that they could get along. Saxby met two senior members of the team, Jamie Urquhart and Tudor Brown, at Heathrow Airport as Tesler, who the pair had driven down from Cambridge, shook hands and dashed off to board Concorde back to the US.

Both Acorn and Apple were content and offered Saxby the job. But before saying yes, he wished to know if he would be accepted by those he would work with closest. And after ES2, he knew he needed to keep costs down and wondered if the dozen people he was being presented with fulfilled the start-up remit.

---

The Rose & Crown was neutral territory, lying between Acorn's base in Cambridge, and Saxby's home in Maidenhead in Berkshire, which he shared with his wife Patti, a teacher, and their two children. As someone who travelled a lot, he already had a mobile phone, but none of the Arm team did to call ahead and warn him they would be a little late.

It turned out to be a cordial night. As they finished their drinks, Saxby set the team a task. He wanted to gauge their opinion of how Arm was placed to succeed. Led by Urquhart, Saxby requested an assessment of Arm's strengths, weaknesses, opportunities and threats: what's known as a SWOT analysis.

It was something to discuss next time they met because the team accepted him unanimously. Saxby joined the company full-time on 18 February 1991 as president and chief executive officer, but in practice was involved straight away as he worked through his ES2 notice. When he saw the potential in something, he was all in.

## Harsh Realities

Compared to the jubilant press notice that had announced the creation of Arm in November 1990, a second, internal, document prepared soon after was far less celebratory. On 18 December, at 10 o'clock in the morning, Arm's engineers downed tools and gathered in Acorn's Cherry Hinton conference room. Over the next hour, they compiled the SWOT analysis that Robin Saxby had requested.

The single typed page that emerged was a carefully balanced study of the pros and cons: not overly negative, not overly arrogant. For the 12 that were being cast out from the relative safety of Acorn it was the moment that reality bit.

Headed 'Arm LTD SWOT', with 'confidential' handwritten in the top-right corner, it set out the new company's advantages of its basic technology and established team, who were variously described as 'flexible, responsive, dynamic, successful (so far), enthusiastic' and with 'extensive systems experience'. Against the pluses were noted several undeniable minuses: a poor commercial starting point, limited resources and a reliance on third-party support.

Of course, the document highlighted many opportunities, both in emerging technological fields and geographies. The 12 could never have known how big some of those opportunities would become, stretching out decades. Yet alongside 'portables' (the closest reference made to mobile phones in the document) and 'embedded control' (essentially any use for microchips beyond the world of personal computers), the immediate threats were clearer. Arm currently had negligible customers, no control over its income and significant rivals. Most revealing of all, next to the word 'patents', one of the group had jotted down, in slanting capitals: 'own none'.

Saxby knew all this prior to reviewing the document. The combination of the ideas incubated by Acorn with a little money from Apple was enough to found Arm. And because the Arm processor was likely to fit into larger chip systems – as one element among several – the obvious route was to license its designs to others to make up and earn royalty income for every device containing its work. But from there Saxby was keen to diverge from the business plan its investors had in mind. 'He threw it in the garbage can,' said Advent's John Berylson.

If Arm was going to survive it needed to go further and faster with its mix of licensing and royalty income than Acorn and Apple envisaged. Chief among Saxby's aims was to make the low-powered Arm chip a global standard, an unusually ambitious target for a British company when many much larger domestic peers preached modesty. And rather than battling Intel in the PC space it had grown to dominate, Saxby was eyeing everything else. To make inroads into embedded devices meant creating a standard set of products with wide appeal and then fanning out to find customers in different categories and geographies, just as he had learned at Motorola.

Saxby's thinking was timely. The relentless appetite for cash that drove plucky also-rans like Inmos into the arms of buyers was forcing many in the microchip industry to reconsider their focus. Soaring interest rates, rising inflation and, in the US, the crisis that struck many savings and loan associations, contributed to a widespread downturn.

Such was the capital intensity of the industry, by this time companies realised they could not do everything. Foundries such as Taiwan Semiconductor Manufacturing Company (TSMC) were beginning to offer alternative manufacturing options. The software used to design chips and the machinery needed to etch those designs onto silicon wafers were becoming the preserve of specialists too.

For the same reasons, rising cost and complexity, the chip companies could no longer be bothered to keep and manage the basic components of design that would not materially differentiate their end products.

Efforts to spin out these 'cell libraries' were underway. Compass Design Automation became a wholly owned subsidiary of Arm's production partner VLSI in March 1991; what became Artisan Components, which would eventually be bought by Arm, was founded in the same year.

The sole purpose of these firms was to maintain a record of ideas and update processes as manufacturing methods changed. They then licensed this so-called 'physical' intellectual property (IP) – as pre-designed blocks covering specific processes such as memory – back to industry players that wanted to incorporate it into their own designs.

Electronics licensing was not new, as the Japanese-made, American-powered transistor radios of the 1950s and 1960s attested. But now the industry was entering a new phase. Convenience, partnership and specialisation: the biggest manufacturers were open to paying a small royalty for something that would otherwise have taken many millions of dollars to develop themselves. Ideas flew across the globe, just as the physical stock of chips had done for years – and all without too much fuss from political leaders. Perhaps this was the moment a company that made nothing could make a difference.

———

Before addressing that opportunity, Saxby had to keep his investors satisfied. Arm was contractually obliged to produce a new floating-point accelerator (FPA) chip for Acorn that would improve the performance of high-end workstations, a market segment it was keen to pursue. Apple just wanted the Arm600 chip for what became the Newton and sent Allen Baum, an engineer who incidentally went along to the first Homebrew Computer Club with Steve Wozniak, to Cambridge to keep communication lines open. Neither product generated income for Arm, although some consulting work followed, and there was negligible demand externally for either of these specialist chips.

Saxby hadn't for a second believed the forecasts supplied by Acorn and Apple for how many chips they were going to buy and what that meant for his income. He knew he had to look beyond both of them. His own calculation was that if Arm had any chance of becoming a market leader, creating a design standard based on the reduced instruction set computer (RISC) methodology, it would have to have progressed to the point where a spectacular 200 million chips were sold in the year 2000, less than a decade hence.[10]

Along the way, if it was to survive, Arm needed to keep costs low. The earlier business plan suggested hiring specialists in sales and marketing and that another fundraising would not be far off. Saxby rejected both ideas, handing the sales brief to Jamie Urquhart and marketing to Mike Muller, with Tudor Brown as engineering director completing the senior trio. The unique properties of the Arm processor were born out of parsimony. There was to be no generous spending on its search for a global audience either, which kept the team tight, delegation frequent and flight schedule relentless.

The financial constraints extended to Arm's headquarters. In keeping with its image as a high-tech start-up, Malcolm Bird had lined up a move to Vision Park in the village of Histon, to the north of Cambridge, where similar businesses were clustering. But Saxby would not stomach the expense and asked for a cheaper alternative to be found.

That job fell to Acorn's David Lowdell, an unsung hero who provided administrative support as Arm was established. When Harvey's Barn was identified, Saxby secured good-quality wooden desks for the price of plastic ones by letting the furniture company photograph them in the 17th-century barn setting for its sales catalogue. A custom-built boardroom table was won on a coin toss and the team installed their own IT network.

Up the barn staircase with acorn-shaped carvings that were an eerie echo of the firm's genesis, Saxby installed himself in the loft beneath centuries-old beams, while the engineers worked below, their desks encircling the barn's main support pillars where farm supplies were once stacked high.[11] Curious American and Japanese visitors lapped up the quirkiness and didn't mind the detour eight miles outside Cambridge. They were guided by directions in the company's first corporate brochure, which advised: 'Turn right in front of the Black Horse and find Arm Ltd on the right past the church.'[12] It was a strangely idyllic spot for a firm that was meant to be courting some of the world's largest technology companies. Staff barbecues were held in the adjacent field.

With an absence of customers, the important thing was to reach out into the world. It wasn't the easiest time to launch. The semiconductor market had rebounded since the mid-1980s slump caused by the dumping of memory chips but Japan, so vital to the electronics industry, saw the bursting of its economic bubble after a spell of over-investment. The team had to hustle harder than they had ever done at Acorn, where their designs had a natural home in the computers they sold.

'We're mere Davids against a lot of Goliaths, but the little guys can still win if they're sufficiently creative,' Saxby told the late spring 1991 edition of Acorn's newsletter, which dubbed him 'Roamin' Robin' for his extensive business travelling. All those overseas trips while at Motorola had not crimped his ambitions to see more of the world. Saxby said he aspired to follow in the footsteps of Monty Python member turned travel broadcaster Michael Palin. But first he had a job to do.

––––––––––

Saxby's sales techniques were already legendary. At Motorola, he managed to get himself invited to a customer's wedding two months after first meeting, and he pursued a similar closeness with Arm's prospects. On hearing that one Japanese businessman was a Monty Python fan, he moved a Tokyo meeting into the hotel corridor for a parade of 'silly walks' to break the ice.[13]

On another occasion, after signing a contract with Sharp Semiconductors for which he had travelled to Japan numerous times in a year, he celebrated with the visiting boss over dinner at his local Chinese restaurant, China Diner in Beaconsfield, with his wife Patti and her mother who happened to be staying with them. Eccentric, relentless, but ultimately highly effective.

Before landing Sharp, the first new contract Arm had in its sights was Plessey. Saxby needed proof of concept that a company other than VLSI could manufacture its designs and Apple was eager to line up a second supplier of chips for the Newton.

Plessey fitted the bill. The long-standing electronics, defence and telecommunications company was by far the UK's top chip manufacturer, having absorbed the activities of Ferranti, the chipmaker that had supplied Acorn – although it was still modest by international standards. Its sites in Plymouth and Swindon were close enough for Arm engineers to be on hand to help with any glitches.

A deal had been agreed and Arm had transferred its designs even before the contract was signed. Saxby placed his trust in an old Motorola colleague, Doug Dunn, who used to run the East Kilbride plant. But a phone call in summer 1991 brought bad news.

What was now GEC Plessey Semiconductors had been acquired a few years earlier. Dunn explained patiently in flat Yorkshire vowels that his conglomerate owners were looking closely at the deal and signing would be delayed. So would payment.

Dunn, who had stayed on under the General Electric Company (GEC), had become one of 80 managing directors reachable by the legendary Lord Weinstock at the touch of the button from a giant telephone console that sat on the industrialist's desk in a dingy office just off London's Park Lane. Weinstock was a giant of British business, relentless in his ambitions that went from shipbuilding to Avery weighing scales, but also careful in his approach, with piles of cash kept in reserve. To his lieutenants he likened his painstakingly constructed conglomerate to the Nimrod surveillance aircraft that GEC also developed – a collection of 100,000 rivets flying in loose formation.

Dunn had become used to wrestling over budgets and had convinced Weinstock there was no point in keeping a semiconductor arm if he wasn't going to invest in it. But GEC's problem with Arm was not the technology. 'They didn't know an embedded RISC processor from a toothbrush,' Dunn said. Their questions were about whether Arm would survive and had enough resources to continue development.

In truth, it was touch and go. The 12 founders were contractually due a pay rise that first summer, and Saxby had to stand up before them in Harvey's Barn and take them through the cash-flow situation. In the end, he gave them backdated bonus cheques. By Christmas, GEC supported Dunn, who thought Arm was a risk worth taking. The team returned to work in January 1992 to find an extra payment waiting for them. Arm was hanging on – for now.

Recalling the vigorous spending at ES2, Saxby was determined to keep a lid on outgoings. Pay freezes were imposed, as was Saxby's Law, which insisted that spending of more than £10,000 needed justifying in a one-page submission of around 600 words.

Arm raised £650,000 from Nippon Investment and Finance as the byproduct of frequent, awareness-raising trips to Japan, but Saxby regarded those funds as only for dire emergencies. More immediately useful was research funding from the European Union's Open Microprocessor systems Initiative (OMI) which was seeking to create broad technology standards.

Arm staff were understanding of the crunch, hoping that share options they had been awarded would be worth something at some point.

The team also had to cope with tragedy. At Acorn, Alasdair Thomas had slotted in under Steve Furber as a careful, solid engineer who designed the Arm3 chip, a revolutionary advance for its time. He was an easy choice to join the Arm spin-off and was lined up as the lead designer of Arm7, another technological step forward.

Thomas – always Al to his colleagues – was one of the youngest in the office and often took on the role of sidekick. He was very funny and often had colleagues in stitches. On his birthday, Tudor Brown received a cassette tape of silly voices recorded by Thomas. 'He could have been Rowan Atkinson,' Brown said, likening him to the Mr Bean and Blackadder comic actor. He could hold his own on a bigger stage too, as seen in a presentation he gave to Apple staff in California alongside Mike Muller, which is still on YouTube.

Thomas didn't have a lot of friends outside work and he struggled with an acne condition that affected his confidence. One day in August 1992, as interest in the Apple Newton and its suppliers stepped up, he was in the office to talk to a German journalist about the advances Arm was making. Thomas took his own life the following day.

'Here was this guy laughing around us, a key member of the team, and we didn't realise that he was so depressed,' Brown said.

Thomas left a note for colleagues to say he was sorry. The funeral was held in Cambridge, followed by a wake in Harvey's Barn.

# Chapter 5

# NOKIA'S MAD PHONE SETS THE STANDARD

## A Japanese Puzzle

Just an hour north of Tokyo, the prefecture of Nagano was the perfect antidote to Japan's always-on, neon-lit urban sprawl. Among its mountains, farmland and abundant wildlife, visitors could tour the historic five-tower Matsumoto Castle, watch snow monkeys at play and relax in hot springs.

One of Arm's engineers, Dave Jaggar, was on his way there in January 1994 for another reason. Together with his bosses, Robin Saxby and Mike Muller, they had tacked a few days' skiing on to their latest business trip to sell Arm's technology around the world. Yet as the Shinkansen sped north-east that Saturday morning from Nara, which had been their base for a week of meetings with Japan's industrial leaders – Toshiba, Toyota, Ricoh and more – Jaggar took in little of the greenery unfurling outside his window.

The New Zealander was pondering a problem that had once again announced itself, this time in a meeting with Nintendo. The video games console maker was a great prospect for Arm, and already a customer of its licensee Sharp. After enjoying considerable success with the Nintendo Entertainment System and the handheld Gameboy, which introduced the Mario and Zelda characters into millions of homes, now it was developing a 64-bit games console.

Sharp was proving to be an enthusiastic introducer and wanted to use Arm designs for Nintendo's latest ROM cartridges – memory cards that stored video. But Nintendo was resistant because when it tried out Arm's architecture it found it was using too much program memory.

As process technologies shrank, the Arm7 chip was designed as a step up from earlier iterations of the instruction set. Fast, power-efficient and small, it clung on to all the plus points that originally excited Apple for its Newton device and launched Arm as an independent company. The exception was with what programmers termed as 'code density'.

Jaggar, who joined Arm only a few months after Saxby and the 12 engineers set up shop in Harvey's Barn, had heard this complaint several times by now. RISC processors worked well by breaking tasks down into a handful of simple instructions to boost performance, but their Achilles heel was memory, where they typically used 30 per cent more than CISC chips.

Adding in memory added cost: a 30 per cent increase in memory pushed the price of a chip up by 20 per cent. At that moment, those extra costs were unavoidable. Compared to the 8-bit and 16-bit code used by the older processors Arm was in the market trying to supplant, its programs were 50 per cent larger.

Despite such early promise, it was a revelation that threatened to stop Arm in its tracks. Jaggar made the bold suggestion that the company needed to junk the Arm architecture and start again. Unsurprisingly, Saxby and colleagues, who had spent several years evangelising its potential, were not keen on such a drastic about-turn. But the short, outspoken Jaggar wasn't shy in airing his views.

What he didn't realise was the solution he was sketching out would do more than simply win Arm some gaming supply contracts. What the company at that time viewed as a far smaller market opportunity, mobile communications, was about to open up to it in astonishing fashion. Solving the problems of handset weight, battery life and cost that so far kept cellphones as a niche, businessman's preserve involved a journey that took in one of the oldest chipmakers, one of its newest customers, the baking heat of Houston and the frozen north of Finland. If it wanted to become a global design standard, Arm had to be prepared to travel.

––––––––––

Jaggar had encountered the BBC Micro at Shirley Boys' High School in Christchurch, New Zealand, and was captivated. By the time the Acorn Archimedes made it to him in late 1987 he was working at the school and several others as a computer technician.

Jaggar's fascination inspired him to dig deeper into how these devices worked. The resulting master's thesis, prepared at Christchurch's University of Canterbury, was entitled 'A Performance Study of the Acorn RISC Machine', so when it came time to find a job, there was only one place to go.

Steve Furber was already professor of computer engineering at the University of Manchester by that time but he forwarded Jaggar's inquiry on to Arm. 'In some ways you may know more about the chip than we do,' Jaggar recalled Furber saying to him when they first met in Cambridge. By June 1991, he was working there among the clever but shy band who introduced him to warm English beer. Saxby even moved into Jaggar's flat from Monday to Friday because it was easier than driving back to his family in Maidenhead every night.

Jaggar saw Arm's undoubted strengths – but also feared its weaknesses. The team was fully qualified but lacked some of the computer architecture expertise he knew that the RISC rival MIPS Computer Systems had on board. 'On the one hand, I felt like I was sitting cross-legged at the foot of the gods,' Jaggar said. 'And on the other hand, I quickly realised that actually the formal training was missing.'

His thesis had touched on the trade-offs between RISC simplicity and the looming code-density problem, but only for high-performance workstations, not the embedded uses Arm was pursuing. Nevertheless, travelling to Nagano, a thought occurred.

If only Arm could ditch the existing 32-bit architecture some of the time to save on memory – but not all the time. As Jaggar knew, some computer code needed to run fast, but most didn't. What he sketched out was a method of using two instruction sets within one chip. Appropriately enough for his location, the analogy he used was the two ways of writing Japanese. There was either the phonetic lettering system of *hiragana*, or there was *kanji*, which condensed two or three characters into one, based on Chinese script. He spied a trade-off.

The idea was one he would have the chance to share sooner than he thought. Shortly after his brief ski break, Jaggar swapped the snow-capped peaks of Japan for the icy north of Finland, home to Nokia.

## Finland's Big Call

When Kaarina Suonio picked up the telephone to Finland's former prime minister Harri Holkeri on 1 July 1991 she was making history. The deputy mayor of Tampere, the country's second biggest city, spoke for just over three minutes with Holkeri, who wished her 'good summer' from his car phone. The conversation was the first to be carried over GSM (which stood for Groupe Spécial Mobile, and later Global System for Mobile Communications), a new European technology standard for digital cellular networks. Holkeri remarked that the reception was so clear it was like talking to 'someone in the next room'.[1]

For that, he had Nokia to thank. Finland's largest company had made great strides with mobile technology since launching its first 'portable' product, the Senator, a handset attached to a bulky suitcase battery, nine years earlier. Having supplied the equipment for the maiden GSM call that garnered international attention, it was aiming to build out a business based on it.

While the world was still falling in love with personal computers in the 1980s, in some markets, and led by a handful of companies, heavy experimentation with mobiles had begun. Although a first cellular call had been made by Motorola engineers in 1974, it was the Nordics, not the United States, that took the initiative.

From 1981 in Sweden and Norway and 1982 in Finland and Denmark, the Nordic Mobile Telephone Service (NMT) allowed customers to make calls anywhere across the sparsely populated region using a shared standard and common 450-megahertz frequency band. The analogue mobile telephone network had been in development for more than a decade by four neighbouring countries that were used to working together.

The service quickly found a market, attracting 110,000 subscribers in the Nordics by 1985. By 1988, Norway had the world's highest density of mobile phones with 33 per 1,000 inhabitants, followed closely by Sweden and Iceland (which joined NMT later). This was despite mobile phones weighing anything up to 15kg at first because of the battery that was needed to support the energy-sapping electronics and radio unit. As the handset shrank, the price tag initially grew, limiting consumer appeal.[2]

Given their early experience, the Nordic nations and their two flagship telecoms equipment makers, Nokia and Sweden's Ericsson, heavily influenced the second-generation (2G) standard for mobile telecommunications. Agreed by a memorandum of understanding signed in Copenhagen in 1987, the GSM standard would operate across most European countries and did away with competing proprietary systems controlled by separate manufacturers. It also marked a shift to digital, so that voice calls were converted into numbers and transmitted via radio waves, taking up less room on a network than the same conversation in analogue.

---

The leader of Nokia's mobile-phone division since 1990, Jorma Ollila, saw the potential. He had been marked out for high office ever since his headmaster recommended him for a scholarship at the Atlantic College, a Welsh boarding school for future leaders set up by Kurt Hahn, the German founder of the Outward Bound movement. Together with time spent in Citibank's London office, the bespectacled Ollila brought with him an international perspective and a hard-to-trace accent when he joined Nokia in 1985.[3]

Mobile phones constituted one of several high-tech diversifications for a company whose name was a reminder of its origins as a timber mill based on the banks of the River Nokianvirta in southern Finland. Despite its home, a relative corporate backwater on the world stage with a population of just 5 million people, Nokia had already led the mobile-phone market in 1987. But as manufacturing volumes rose, it buckled under the pressure, began losing money and ceded top spot to Motorola.

GSM gave it fresh impetus, just as Nokia was in dire need of new products it could sell into new markets. In the same year that Holkeri made his famous phone call and Arm's launch team was finding its feet in Harvey's Barn, the Soviet Union collapsed. The political superpower across the border had been the destination for a quarter of Finnish exports. Ollila saw his domestic economy battered and watched the unemployment rate leap to 20 per cent.

The following year, 1992, he was appointed chief executive of Nokia with a mandate to take bold action. Ollila applied a steely resolve to focus down on the mobile-phone division, which saw him complete Nokia's transformation from an industrial conglomerate. Legacy assets – the production of tyres, power cables and televisions – were offloaded. Ollila knew Nokia could shift many more phones if it could capitalise on the economies of scale that GSM offered – but to do so it needed to find the right partners.

Nokia's introduction to Arm took place far below the CEO's pay grade and had initially been presented as a joke. In February 1991 the company had strengthened its position with the £34m acquisition of Technophone, the UK's only cellular phone maker, to make it a clear second in the industry behind Motorola.

When Technophone joined Nokia, some of the staff at its base in Camberley, Surrey, were amused to discover there was at least one Brit already in the business. Craig Livingstone was a software engineer based in Oulu, the far northern reaches of Finland, even though he had been born in Middlesbrough, a town in the northeast of England. In a Finn-heavy company, that made him stick out as he explored which microprocessors Nokia might use as it planned to scale up mobile-phone production.

Appreciating his professional interest, one of the Camberley crew, an enthusiastic Acorn Archimedes user who had a rudimentary grasp of the Nordic lifestyle, sent Livingstone something in the internal mail. When the envelope arrived, Livingstone pulled out Arm's first customer brochure, which featured a picture of Harvey's Barn on the back page. A Post-It note stuck on top declared: 'Thought you might like this – it looks like a sauna!' Arm's quirky headquarters continued to pay dividends.

## The Worst Deal Ever

In the decades that followed Jack Kilby's 1958 breakthrough, his company, Texas Instruments, had put it to good use. It parlayed integrated circuits into missiles, radar, memory, calculators and computers to make it the world's biggest maker of microchips for a time. But by the 1980s it was ailing.

With the benefit of hindsight, critics said the go-getting Pat Haggerty's vision of making money from equipment, not the components they contained, had squandered TI's advantage in semiconductors. The group had too many underperforming businesses in its portfolio, and one critical opportunity, the PC microprocessor market, was lost to Intel.

The answer was for TI to throw its weight behind specialist chips that could handle graphics or storage. A third type, dedicated to communications, owed some of its genes to the Speak & Spell electronic toy that TI had produced in 1978, featuring an integrated circuit that mathematically modelled the human voice.

Developed in Houston and introduced in 1982 for commercial use, its digital signal processor (DSP) chip, the TM320, translated sound and images from analogue into digital. TI's first DSP customer put them in analogue repeaters in underseas cables, which were used to extend the distance a radio signal could travel. Then came IBM disk drives. But the company needed to find some high-volume applications to make a difference to its bottom line. Some TI staff thought the DSP had a future in PCs, but others were already turning their attention to the developing mobile market – and particularly would-be customers in Europe.

TI had always had an international mindset, forging alliances around the world. One such cooperation agreement was signed in 1987 that traded Ericsson's telecom systems knowledge with TI's semiconductor expertise. That grew into a cross-licensing agreement to develop applications together. Ericsson built a wafer fab at Kista, just north of the Swedish capital Stockholm, with engineers travelling to TI in Dallas to learn about manufacturing. During nights out to Stockholm's grand, waterfront opera house, TI's semiconductor chief Wally Rhines convinced Ericsson chief executive Lars Ramqvist to use its DSP design for the company's first digital cellphone.

A similar charm offensive was going on with Nokia in neighbouring Finland. An alliance was marked by a celebratory crayfish dinner in Helsinki, an event attended by TI's Thomas Engibous, who would move up to succeed Rhines in 1993, and Gilles Delfassy, an expansive Frenchman who was responsible for TI's DSP push in Europe.

Delfassy was based in Nice, handily a half-hour drive from the European Telecom Standard Institute (ETSI), which was developing the GSM standard at its base on the Sophia Antipolis technology park. There was still doubt over whether the technology would prove to be too expensive to roll out but TI thought it was worth pursuing.

The company designed a DSP into a cellphone for Ericsson but it was clear costs needed to be driven down if it was to work commercially. Its Nordic rival Nokia was bolder.

There were three main elements to the circuitry required to power a mobile phone: a microcontroller, which acted as the device's 'brain', a customised application-specific integrated circuit (ASIC) and a DSP. TI tried, and failed, to win the contract to supply the first generation of Nokia's digital phones with any of them, but for the second generation it displaced AT&T to contribute the DSP, vindicating its strategy.

For the third generation, the first that would adopt the GSM standard, the two companies retreated to the Finnish forest to think. During a three-day offsite the top 10 most relevant technical staff from TI and Nokia worked, ate, drank and took saunas together. When they emerged, they had drawn up a roadmap for the future of mobile telephony as they saw it.

By this time they sensed that digital cellular technology had the potential to take over the world, but handsets had to be a good degree smaller and, to become indispensable to consumers, they had to hold their battery charge for several days. Lower power consumption was essential.

Nokia calculated what it needed to do to regain market leadership from Motorola. One potentially game-changing solution was to integrate the three separate components – the microcontroller, ASIC and DSP – onto a single 'baseband' chip that would manage all the phone's radio functions. It was an adventurous move; some thought foolhardy. No wonder the several hundred designers deployed on each side came to know the project they were assigned to as 'the MAD phone', an acronym for microcontroller, ASIC and DSP.

'We did the maths backwards and concluded that we couldn't make enough phones in our factories if we didn't reduce the component count,' said Tommi Uhari, who became programme manager of the single-chip project. The phone had to become cheaper and simple enough to be stamped out reliably millions of times.

The chipmaker TI rethought everything for its partner. Hitachi had supplied the microcontroller for the second-generation phones but this time power usage was so critical that TI dropped the Japanese firm. It didn't even think its own microcontroller design was good enough.

Instead, it took a risk on a promising UK start-up that was beginning to make itself known for its 'very brilliant architecture' according to Delfassy, 'but honestly it was struggling to get traction in the marketplace'. TI's interest in Arm had been pricked when Mike Muller set out its technological advantages at the Microprocessor Forum, held in October 1992 at San Francisco's Hyatt Regency Hotel.

But mindful of TI's past setbacks and the markets it had lost ground in, when asked to sign a licence with Arm, Rhines exclaimed, 'We can't manufacture anything, now you're telling me we can't design anything either?'

---

Arm's early dialogue with Nokia brokered by Craig Livingstone had yielded a decent relationship but no revenue to date. Pete Magowan, a product marketer who joined Arm from ST Microelectronics in 1992, became a regular visitor to Nokia's base in Oulu, a small city 100 miles outside the Arctic Circle.

At first engineers and salesmen that gathered there from all over the world found it a novel assignment. Visitors stayed in Oulu's main hotel that looked out on the town square and sometimes joined the local disco on Thursday nights. But the lack of daylight, which meant arriving for meetings and departing again in the dark, and curious snacks, such as salty liquorice candy and chocolate cheese, meant that quickly paled.

Because neither Nokia nor Arm made their own chips, Arm needed to be licensed through a manufacturer that did. When TI took out a licence in May 1993 it was a ringing endorsement from one of the world's biggest semiconductor makers, but also one of the oldest that reached back in time to the industry's earliest days.

However, finding its way into Nokia's phones was not yet a given – especially when TI and Nokia discovered the same code-density problem that Nintendo had raised with Dave Jaggar. Fortunately, at a meeting in Oulu in February 1994, fresh from the Japanese ski slopes, Jaggar put forward what he thought was a solution.

The Arm7 design could be adapted by adding a new 16-bit instruction set within the 32-bit architecture to carry out the most common tasks. It would produce far less code, solving the memory problem. Nokia was interested in principle, so Jaggar carried on sketching on a drinks napkin on the flight back from Finland.

With some refinement, what emerged was the ARM7TDMI, incorporating the Thumb extension, so-called because it was the 'useful bit on the end of an Arm'. Programs compiled with the Thumb add-on were found to be 70 per cent more compact and ran approximately 50 per cent faster than 8-bit or 16-bit wide memory. That meant better performance, lower cost and lower power consumption, delivering precisely what Arm had been marketing in the first place. Nokia tested the new solution by supplying it with a chunk of GSM source code. It was impressed by the result.

Amazingly, the innovation did not go down well with everyone back in Cambridge. In an email sent to Jaggar's superiors – but not him – on 13 June 1994, Sophie Wilson, the co-creator of the Arm architecture with Steve Furber, did not mince their words in reaction to their work being unpicked.

Wilson warned Jaggar's superiors that Thumb would be a disaster for Arm's architecture – a backward step.

The grumbles did not hinder progress. After 15 months of work, a prototype of the MAD phone, including the Arm microcontroller core that employed Thumb for the first time, was delivered in April 1995. On a conference call one day, Timo Mukari, Nokia's software line manager, told Arm's Magowan, 'We think we might ship quite a lot of your processors.'

To make that happen, the price that TI, as Nokia's supplier, would pay Arm was fixed early on. 'We told them: if we select you, you will enter the cellular market big time,' said Delfassy. As a result, he added, 'We want very, very favourable terms.'

For every device that featured Arm's designs, TI drove down the royalty fee it would pay to a fraction of a cent per unit. And, with an eye on the future, if several Arm cores featured in the same chip, TI stipulated it would only pay a single royalty. Firmly emphasising who was in the driving seat, the deal was also struck in perpetuity.

'Apple have played hardball with us on royalties,' one former Arm director said, 'but nobody has ever succeeded in getting anywhere near to TI levels.'

That didn't seem to matter when volumes smashed all expectations. In 1997, Nokia sold 21 million mobile phones, giving it a 21 per cent market share. In 1998, that rose to 41 million handsets and the company's share price tripled. It overtook Motorola to become the world's number one in a market that had grown overall by 51 per cent.[4]

Its 6110 mobile phone had been announced in December 1997, boasting five hours of talk-time, a one-touch voicemail button and 35 ringtones for its initial target market of businesspeople to choose from. Consumers might recall it was the first cellphone to be installed with the addictive Snake game but, almost five years after the TI licence had been signed, it was also the first with a processor based on Arm designs.

Arm's contribution enabled Nokia to set a new benchmark for its own chipsets. The component count in its new device was halved and the cost and power consumption beat the rest of the industry by a wide margin. That meant the phone could get by with a smaller battery and was therefore lighter, at 137g, and truly pocket-sized compared to the bricks of the past.

But it did more than that, effectively creating a new category of device that Nokia dominated for more than a decade. Since the development of the integrated circuit there had been radios, calculators, laptop computers and handheld gaming devices, and, for the business market, pagers, PDAs (personal digital assistants) and hefty mobile handsets. Fulfilling the long-held dream of portable electronics for the masses – a gadget that people would not leave home without – was the basic, lightweight mobile phone.

Nokia couldn't have done it unless the silicon had stepped up. By 2000, TI derived 85 per cent of its revenues from DSPs and the analogue chips they work with in tandem to convert information and then compress it. The processor also changed Arm future. Two decades on, the ARM7TDMI design would still be selling several hundred million times every year, used in devices far removed from mobile telephony.

Nintendo eventually took the ARM7TDMI design too but was slower to move because it relied on third parties to write its games. In comparison, Nokia was in control of its own software. Ericsson also lagged. Initially, it didn't want to adopt the same integrated baseband chip as Nokia for fear TI would build a monopoly position in mobile like that forged by Intel in the PC market. When it changed its mind, it never really caught up. By 2001, Ericsson had sounded the retreat, rolling its handset business into a joint venture with Sony and vowing to concentrate on the network equipment business instead.

As chips became more and more complicated, the industry woke up to the possibilities of using Arm. No longer did chipmakers need their own in-house microprocessor design team when they could simply license it in from someone else – at a modest price. The TI endorsement ensured the dam burst. There would be no more struggles to meet the wage bill. Notably, Samsung signed up as a licensee in 1994 on terms more favourable to Arm and after only a handful of meetings.

On one level, Arm's TI deal was the worst it ever struck. On another, it was undoubtedly the best.

## Chapter 6

# AS ARM CASHES IN, APPLE CASHES OUT

## The Return

When Steve Jobs returned to Apple in January 1997 the company he co-founded was in desperate need of new ideas and new money.

Jobs emerged on stage at the San Francisco Marriott to flashing light bulbs and a 30-second standing ovation, having been introduced during the marathon Macworld event by Apple's chief executive Gil Amelio. Dressed in black cardigan, white Nehru-collared shirt and baggy trousers, a few remarks about his 'mission', delivered with confidence and clarity, were lapped up by the faithful developers and customers that filled the room.

'What we want to try to do is to provide relevant, compelling solutions that customers can only get from Apple, right?' Jobs said to more clapping, his tone marking a welcome break from Amelio's disjointed performance. 'That's what we want to do because if we can't figure out how to do that then I think there's a lot of other options for people to buy their computers from – and we might not sell enough computers.'[1]

Apple was in dire straits. In the year to September 1996, it had tumbled to an $816m net loss on sales 11 per cent lower at $9.8bn, precisely because it was not selling enough Macintosh computers. In its annual report, the company put the decline down to 'customer concerns regarding the company's strategic direction, financial condition, and future prospects', as well as delays to its PowerBook 5300 product, the first with chips from its PowerPC alliance with IBM and Motorola, which had been dogged by mishaps including flaming batteries and suspect software.[2]

The outlook was no better, with net sales set to 'remain below the level of the prior year's comparable periods through at least the first quarter of 1997, if not later', Apple added. A major cost-cutting drive was in progress and, to keep its lenders happy, the company was eking out cash where it could. Some $145m had been scraped together during the period from the sale of equity investments, a data centre and a manufacturing plant.

The crisis explained Jobs' return. Eager to reinvigorate Apple's range with a new operating system, Amelio, a turnaround expert who had only taken the helm in February 1996, knew he had to buy something in when what was developed in-house turned out not to be fit for purpose. His choice boiled down to acquiring Be, a venture led by Jean-Louis Gassée, Apple's former head of product development and marketing who had set the Newton project in train, or Jobs' NeXT, which had struggled enough to have already given up on making hardware.

Both sides pitched and Amelio consulted. The long-serving Larry Tesler, who had risen to become Apple's chief scientist, favoured NeXT but told him, 'Whatever company you choose, you'll get someone who will take your job away, Steve or Jean-Louis.'[3]

Tesler was spot on. Six months after the San Francisco event – and only seven after Apple said it would pay a generous $429m for NeXT – Amelio was gone. It did not take long for the influence of Jobs, acting as a part-time adviser at first, to percolate through the organisation, impacting key decisions around appointments and strategy.

By the time of the next Macworld event, at Boston's Park Plaza hotel in August 1997, it was clear who was in charge, even though Jobs would not be confirmed as interim CEO for another month and did not become Apple's actual CEO until January 2000.

In Boston, he introduced himself on stage as 'chairman and CEO of Pixar', the animations studio he was officially running, adding that he was one of several people 'pulling together to help Apple get healthy again'.[4]

The most surprising addition to that group of supporters was Bill Gates, the chief executive of Microsoft with whom Apple had been battling for years over patent infringements. The pair had not only resolved their legal differences, but Apple would make Microsoft's Internet Explorer the default browser for the Macintosh.

More importantly in the short term, Jobs had negotiated for Microsoft to make a $150m stock investment in his cash-strapped company. 'We're pleased to be supporting Apple,' Gates said, looming over the stage via satellite link to alternate cheers and boos. 'We think Apple makes a huge contribution to the computer industry.'[5]

Jobs had obtained some cash, but he would need more. And now he had to turn his attention to Apple's product line-up.

## Ahead of Its Time

More than three years prior to Jobs' on-stage return, another champagne bottle joined the lengthening row along the shelf at Harvey's Barn to mark a very different event. On 2 August 1993, the Arm team gathered to toast the launch of the Newton device, as it had become known to all, although the brandished by Apple chief executive John Sculley at '10.30am Boston time' – as someone scribbled on the Moet & Chandon empty – was actually branded as a MessagePad.

There were more drinks at the UK launch. Robin Saxby presented each of his staff with a bottle of Springbank single malt whisky, after a doubting journalist wrote in *Byte* magazine that he would crack open a bottle of the stuff if the Newton was ever shipped.

It had been a long wait for the gadget responsible for Arm's creation. The early announcement of a 'personal digital assistant' (PDA) by Sculley in January 1992, followed by the reveal of a prototype that May, allowed rivals to enter this new market before Apple was ready. And, as he stood on stage at the Boston Symphony Hall holding up the first MessagePad, which measured 11.4cm by 18.4cm (4.5x7.25 inches), weighed less than half a kilo (under a pound) and was priced at a toppy $699, Sculley was himself on borrowed time.

In June 1993, he had been asked to step down as CEO by Apple's board, who were fearful of the company's collapsing share price. By October that year the Newton's great champion would have relinquished his role as chairman too.

Containing a 20-megahertz Arm 610 microprocessor, the MessagePad attracted plenty of attention but little of it good. Its ability to organise messages and meetings and an infrared link to other MessagePads enabling information to be shared was overshadowed by its patchy handwriting software. Using a plastic stylus to jot down notes on screen led to some comic interpretations of what users were trying to record and the MessagePad's ridicule in Doonesbury cartoons and beyond.

The Apple faithful raced to buy it. Some 50,000 Newtons were sold in the first 10 weeks on the market. That was roughly the same rate as when Apple launched the Macintosh: respectable, but a long way short of the blockbuster success it needed. The

Newton operating system was licensed to Sharp, Siemens and Motorola so they could produce their own PDAs, but that did not develop into a significant revenue stream either.

The doubts did not stop Saxby from using the launch as a marketing opportunity for Arm. 'Part of our vision when we set up this company was that Arm chips would be used by everybody in the world. For that to happen we need to become a household name like for example Dolby,' he said, namechecking the US company that licensed its audio technology to numerous consumer electronics firms in an interview with Arm's local TV station Anglia.

'Five years hence I'd like a large percentage of the world to have heard of Arm. I think if that has happened most of the other things will have fallen into place: we'll be a big company, we'll be trading very profitably, we'll have offices around the world,' he added.[6] Saxby was posing in front of an 'Arm powered' logo, the company's short-lived effort to create greater brand awareness. Five years on, Arm had prospered further, but no thanks to 'Arm powered', nor the Newton, which was dead.

---

On his return to Apple in 1997, Steve Jobs' disdain for Sculley's project was clear. 'God gave us ten styluses,' he said, waggling his fingers. 'Let's not invent another.' Jobs suggested to Gil Amelio in a phone call, 'You ought to kill Newton.' Amelio wasn't keen, but the die was cast.[7]

The product was far from being a total failure, especially the upgraded, cheaper models that had followed. The MessagePad 2000, marketed as 'the only handheld computer you can actually use', and the related eMate devices had sold reasonably well since their introduction that March.

On 22 May 1997, Apple announced it would spin off its Newton division into an independent, wholly owned subsidiary. That decision involved moving 170 employees off Apple's Cupertino campus, although Apple would remain a licensee of the technology. Fred Anderson, Apple's chief financial officer, declared that outside investors for Newton Inc could be sought in time or an initial public offering was a possibility. 'We expect significant growth and profit in Newton's first year,' he said.[8]

What sounded like a new beginning was really a protracted end. As Jobs asserted himself over Apple once again, Newton's prospects dimmed. In September that year, just after the Microsoft investment had been secured, he told executives at the newly spun-off company not to bother moving into their new office.

Newton was folded back under the Apple banner with efforts focused on building up the eMate 300 model. But by February 1998 the product line was discontinued so that the company could 'focus all our software development resources on extending the Macintosh operating system', Jobs said.[9]

It was a case of what could have been. With the benefit of hindsight, the product was a crucial staging post for Apple on the road to better things. 'It's fascinating because the Newton was basically the iPhone in concept,' said Tudor Brown, who was Arm's lead designer of the 610 chip. 'But obviously it was far too big and clunky, didn't actually fit in your pocket, and didn't have wireless communication.'

On his retirement in 2012, Brown received a mounted MessagePad complete with the inscription '17 years ahead of its time'. But given its flourishing relationship with TI and Nokia, Arm did not have to wait that long to demonstrate it really did hold the key to the mobile revolution.

## A Failed Takeover

The video screen flickered and the image of Fred Anderson, Apple's chief financial officer, appeared larger than life before the group assembled in the Cambridge meeting room. David Lee, Acorn's managing director, could immediately tell it was not good news.

In spring 1997, videoconferencing was much talked about but a virtual meeting of this kind was still unusual. It was expected to be a final round-up prior to an important trip. A team from Acorn was shortly due to visit Apple's headquarters in Cupertino, California, to rubberstamp a secret agreement between the two companies that had banded together to create Arm seven years earlier.

To the horror of the handful of Arm's senior team that knew about it, that partnership looked to be over. Apple had agreed a price at which Acorn could purchase its stake in Arm. Lehman Brothers, Acorn's financial adviser, had arranged the necessary fundraising. Flight tickets had been bought so hands could be shaken and contracts signed.

And now this. Anderson, hair always well-coiffed and with white teeth gleaming, carefully explained that, with regret, the deal was off, but declined to go into further detail. According to those present, Lee was dumbfounded.

This meeting marked one of the most dramatic moments in a year when relations between Acorn and Arm, corporate parent and child, hit rock bottom. Their diverging fortunes lay at the heart of the tension. Put simply, Arm was rising and Acorn falling. As managing director, Lee was intent on holding the two sides together but Arm's leaders, sensing their potential, were intent on securing full independence.

Always dapper, methodical and with a good sense of humour, Lee had been drafted in to lead Acorn in July 1995 after the abrupt departure of the authoritarian Sam Wauchope, who appeared to regard Arm as merely Acorn's subsidiary, there to serve the parent company's needs. The diehard rugby fan had unauthorised trip to attend the World Cup for several weeks in South Africa, and left shortly afterwards.

Lee was a chartered accountant who had been Olivetti UK's director of finance and administration for several years, having joined the firm in 1981. He described his new role as an 'exciting challenge', adding that he wanted to work with Acorn's team to 'identify and implement new initiatives to ensure the long-term success of the business'.[10]

But the more he saw, the more concerned he must have become. Acorn's financial results for 1995 bore the scars of Wauchope's departing profit warning. The company recorded a £12.3m pretax loss, which had widened significantly from £3.4m in 1994. The proceeds of a £17m rights issue were used partly to eliminate bank borrowing. Acorn's sales of computers and services into schools were suffering and telecoms providers had been slow to trial its digital interactive TV technology.

———————

One venture remained a source of hope and it involved an interesting character from Silicon Valley. Larry Ellison was the well-connected founder of Oracle, the database software firm, a sailing fanatic with bouffant brown hair, designer stubble and boxy suits. The Windows 95 operating system had just launched with great razzmatazz and Ellison wanted to create something that would break Microsoft's stranglehold on personal computing.

On 4 September 1995 he used a speaking slot at the International Data Corporation's European IT Forum in Paris to announce his retort. A PC was a 'ridiculous device', he declared:

overpowered, too expensive, hard to work and increasingly irrelevant.[11] His vision was for what he called a 'network computer' (NC), which Oracle would help to build within a year. For $500, consumers or businesses could buy a far simpler device than the PCs currently on the market. It would serve one purpose: to connect users to the internet.

For its time, it was a bold and exciting idea that got everyone from IBM to Apple thinking. And it represented a potential return to the big time for Acorn – or perhaps a last roll of the dice.

The firm's enterprising Malcolm Bird – who had been the architect of Arm's birth – successfully pitched Acorn's set-top box technology to Ellison on a trip to California fixed by Olivetti, which was still a major shareholder. The agreement for Acorn to design the NC incorporating Oracle software was sealed on Ellison's return trip to Cambridge. 'I've got a presentation in nine weeks or whatever it is and you've got the contract. I want a reference design for the Network Computer,' Bird recalled Ellison saying as he departed the meeting.

However, when it went on sale in August 1996, the device did not live up to the hype. It had been rushed to market and the ARM7500FE-based processor it used seemed miscast and underpowered. Just like the Newton, the NC was ahead of its time. Whereas Arm had bigger opportunities to pursue, Acorn desperately needed a hit. The NC attempted to run web applications instead of installed software, but long before regular internet access was good enough to support the idea. His partners suspected Ellison lost interest early on.

———————

It meant that Lee had little alternative but to put his plan into action. Once an unproven drain on resources, now Arm, which turned a £3.3m pretax profit in 1995, was the jewel in the group's crown and an initial public offering (IPO) was under discussion. But as he watched Saxby recruit extra staff while he was forced to downsize, Lee thought he could acquire Apple's share to save Arm the trouble of floating – and rescue Acorn into the bargain.

Aside from its increasingly international perspective, the younger company had already displaced the old locally. In March 1994, in need of more space, Arm moved back from Harvey's Barn in the bucolic Swaffham Bulbeck to Fulbourn Road in Cherry Hinton, taking over from Acorn the old waterworks where it all began and, later on, the Silver Building to the rear of the site. With no sense of irony, Acorn consolidated its staff on the Histon business park to the north of Cambridge that Saxby had rejected as too expensive.

Lee put the idea of combining the two companies to Chris Rodgers, a fund manager at Schroders and a major Acorn share-holder. Rodgers thought Apple could swap its stake for shares in the combined entity and assumed Arm was worth about £200m. He wrote to Saxby in December 1996 to let him know about what Lee had mooted. Arm rejected the idea.

The days of Acorn and Apple's rivalry were long past. In February 1996 they pooled their operations that sold computers into UK schools into a joint venture entitled Xemplar. Despite Arm's froideur, talks for Acorn to buy out Apple from Arm began late that same year.

From Apple's perspective, the Arm venture was another hangover from the John Sculley era, but its dissolution is not thought to have been a plan that Steve Jobs was closely involved in when he returned in January 1997. Instead, it was Anderson, still keen to drum up sources of cash to repair Apple's balance sheet, who listened carefully to what Lee had to say. Executives recall he was not overwhelmed by Arm's progress. 'I thought you guys were bigger than this,' Anderson muttered on a trip to Cambridge, one remembered.

To explain why the deal failed, insiders point to Larry Tesler. Apple's chief scientist remained on Arm's board and was a great advocate for the business and mentor to Saxby. Why sell now, he asked Anderson, so the theory goes. You could ride out the next six to nine months and receive a better return when Arm goes public.

Tesler was right, and Arm turned out to be worth far more than Schroders' £200m estimate. Unsurprisingly, Lee may not have agreed with Apple's decision. He had succeeded in narrowing Acorn's pretax loss to £6.3m in 1996, but the business was still running dangerously short of options to secure its long-term future.

Many of the Arm team only found out much later how close Apple came to exiting. Who knows what would have happened, who would have left, how Arm's strategy might have changed. It's clear that returning to the Acorn fold after seven years away would have changed the course of Arm's history.

## Nasdaq Calling

Given Arm's global ambitions, there was no reason for its stock-market debut to be limited to the UK either. It made sense to offer shares in an initial public offering (IPO) on the London Stock Exchange (LSE), where Acorn had traded for many years, but Nasdaq was alluring.

The New York-based exchange had grown to become home to a crop of fast-growing technology stocks. Arm didn't need to raise money but it did need to boost its legitimacy with potential customers and show it could cope with the pacy, quarterly reporting cycle favoured on Wall Street.

'We saw being on Nasdaq like being at the Olympics, while being on the LSE was more like being at the Commonwealth Games,' said Jonathan Brooks, who became Arm's finance director in March 1995. 'We wanted to be listed on the main technology stock market with the heavyweights of the tech industry.' And a fancy valuation wouldn't hurt either.

Brooks knew the Arm story intimately because of a long-standing friendship with Saxby. Their families had both lived in the Maidenhead area and got on well. Patti, Saxby's wife, had taught Brooks' youngest son and their families had holidayed together. Before Saxby had taken the Arm job in 1991 Brooks had, for the price of a curry, helped formulate a five-year business plan. Relaxing on a break in Corsica together in 1992, Saxby brought along an Apple Newton device for Brooks to cast an eye over.

When Brooks moved to France for a job with the hotels group Accor they kept in touch. And when Saxby suggested there might be an opening for him at Arm Brooks knew it was time to come home.

On joining, the work to make the company IPO-ready began at once. Arm was eager to loosen ties with its founding shareholders and, first of all, that meant dropping the audit firm they both used – and especially the audit partner they had in common.

The numbers didn't always bear out the ambition. Running complex software on embedded chips was still seen as a niche market. In 1995, Arm didn't sign a new licence deal until June and revenue was hard to come by. And when customer numbers grew in 1996, the partnership model showed signs of strain. It became harder to tell each licensee that Arm would automatically give them a competitive advantage, so Saxby deployed his powers of persuasion to explain that actually it gave them the tools to develop their own advantage.

Anticipating an uptick, Brooks alighted on a model for recognising revenues that better demonstrated to would-be investors that Arm's mix of licensing and royalties could add up to predictable growth quarter after quarter. Royalty income had been slow but Saxby hiked the price of a licence in the early years to compensate. Nevertheless, Brooks began canvassing UK investment banks to find advisers for an IPO, but Larry Tesler told him to go for one of the leading Wall Street names. In late 1996, Morgan Stanley was appointed to the task.

---

Investor appetite for technology stocks was mushrooming. In May 1997, Amazon.com, then just an internet bookseller, made its market debut. Of more significance to Arm was a lesser-known US firm called Rambus, which went public a day before Amazon. Boasting technology to improve the communication between chips, Rambus followed a similar licensing model to Arm and saw its shares more than double in value on their first trading day.

External conditions were positive, but closer to home there were problems. Morgan Stanley had strongly advised Arm to pursue a 'pure play' model, that is, not to get drawn back into Acorn. Acorn had reluctantly sanctioned Arm's IPO preparation in April, but others thought David Lee still harboured ambitions of reversing Arm into Acorn instead. Stood behind him was Gordon Owen, Acorn's chairman, who, after spending 37 years at the international telecoms provider Cable & Wireless, was no stranger to bruising boardroom encounters.

In a note prepared for Arm's board meeting on 23 July 1997, Dhiren Shah, one of the team of Morgan Stanley investment bankers, summarised that after months of not taking a firm view on an Arm IPO, Acorn had admitted it saw value in taking Arm public and acknowledged that the two companies did not need to be merged into one.

But soon after, the Acorn board attempted to slam on the brakes in astonishing fashion. One Saturday morning that summer, Arm's directors each answered the door to a motorbike courier who handed them a letter. Despatched by the City law firm Ashurst Morris Crisp, which was acting for Acorn, it sternly reminded them of their fiduciary duties, confirmed the IPO plans were cancelled and instructed them not to make any public comment about the matter.

It was an emotional time, seen as a betrayal by some. 'The impact of the letter was one of disbelief,' said Arm's Jamie Urquhart. 'Why not just talk to us?' The directors did not write back.

Still, the clash was sufficient to throw into chaos plans for Arm to file its intention to float with US regulators in October 1997. It was frustrating: if Arm didn't crystallise some of the value from the options handed to staff years earlier, Saxby feared the company would lose people and momentum.

---

What changed was the appointment of Stan Boland as Acorn's finance director that autumn. The young numbers man from the ICL computer group was inspired to join Acorn when he saw Larry Ellison hold aloft a new network computer at a conference in London and boast that it was designed in the UK. As soon as he started work, Boland discovered the NC contract with Oracle had been terminated.

Inside the company, he encountered 'an army of very capable, but rather disorganised engineers working on a variety of different projects, some of which were paid for and some which were not paid – a cornucopia of different things'. Boland quickly concluded that Arm was totally different to Acorn. 'It was engaged globally with big players and had enormous potential. It was equally obvious it would be completely retrograde to pull it back into a slightly broken UK company to help fix it.' The stand-off was clearly damaging both sides. In time, Boland brokered peace.

When Arm's IPO planning resumed, it was clear some long-standing partnerships were about to pay off. The announcement of Nokia's Arm-powered 6110 handset in December 1997 could not have been better timed.

Investment bankers wanted more sizzle in Arm's IPO prospectus, so Saxby asked the Nokia chief executive Jorma Ollila for permission to use a picture of the phone to demonstrate a real-world application. On the investor roadshow, the handsets caused excitement, especially among US fund managers, because Europe was streets ahead in mobile communications. It hardly mattered that the Newton had been discontinued by Apple in February 1998.

The 110-page prospectus spelled out Arm's growth curve. Revenues of £9.7m in 1995 had risen to £16.7m in 1996 and £26.6m in 1997. Corresponding net income rose from £1.9m to £2.6m and then £3.4m. By 'creating a strong network of partners' Arm was 'establishing its architecture as a leading RISC processor for use in many high volume embedded microprocessor applications', it set out.

Naturally, Arm's lawyers had also loaded the document with risk factors. Quarterly results could be unpredictable because of the timing of new licences and the time taken for royalties to feed through, it warned. Semiconductor partners 'may come into conflict with each other, creating disincentives to market the Arm architecture aggressively'. And the company's revenues were still heavily concentrated, with its largest licensee accounting for 9.7 per cent of revenues in 1997. At this stage, it wasn't even the biggest player in the RISC microprocessor market. Arm's volumes lagged behind MIPS and Hitachi, and a host of competitors in the embedded segment were detailed.

Although much of Arm's IPO preparation was done with the US in mind, Morgan Stanley confirmed investor interest in the UK, which was traced to Lee Morton and Andrew Monk, two salesmen at the City broker Hoare Govett who had followed Acorn for years and evangelised about Arm during its development. They were sure there would be enough demand to justify a stock listing in London too.

Dual listings were in vogue, so the board pressed ahead with plans on both sides of the Atlantic. Given the management team was UK-based, London took the primary listing. For the UK prospectus, Brooks' team added a 30-page preface to the longer document demanded by Nasdaq. Arm was one of a crop of tech companies dashing to market to capitalise on investor demand. Its business model made it harder to understand than many, but those shareholders that got it would be handsomely rewarded.

## Millionaires

On Friday, 17 April 1998, Arm sold 5.8 million American Depository shares in New York and 3.28 million ordinary shares in London, offering about a quarter of the company to external investors. Arm stock soared 46 per cent higher on Nasdaq on the first day's trading, closing at $42.50. Its shares did almost as well in London, surging from 575p to 820p. It was a stronger start than anyone involved had anticipated, including the employees, who owned 10 per cent of the company between them. It gave Arm a $1bn valuation on day one.[12] Steve Jobs relayed his congratulations in a 30-second phone call with Saxby.

The shares kept rising and the media were captivated by the story. The potential for Arm's chip designs was one thing, but the wealth it was already creating was quite another. In a report entitled 'A Fortune in the Fens' for the BBC's regional news programme *Look East*, Arm was cited as having 33 millionaires within the business, including Saxby, who on paper was valued at £36m.

'You might think we have all made so much money we can all now retire and in fact the opposite is true,' said Saxby, who was pictured touring the now-vacant Harvey's Barn in the film.[13] 'It is a bit like we have just joined the Premier League,' the Liverpool FC fan said. 'The spotlight is on us, so there is work to be done.'

———

Amid all the buyers, Apple was a seller. Shareholders had split a £5m dividend before the IPO, and then it offloaded 18.9 per cent of its shares, recognising in its accounts a gain of $24m. The company banked another $37.5m when it sold more in October 1998, reducing its interest in Arm to 19.7 per cent. In its 1998 10-K annual report, filed just before Christmas that year, Apple declared it 'no longer has significant influence over the management or operating policies of Arm'.

Board representation for Apple and Acorn was reduced from two seats prior to the listing to one seat each. On 15 January 1999, Apple's remaining representative, the group's treasurer Gary Wipfler, resigned, and the company subsequently let Arm know it no longer wanted a board seat. However, Larry Tesler, who departed Arm when he left Apple, had rejoined the board in March 1998 in an independent capacity. He would stay on until 2004.

Apple's selling continued: it raised a total of $245m from stock disposals in its 1999 financial year, $372m in 2000, then $176m in 2001 and $21m in 2002. By the time Apple sold its final handful of Arm shares during 2003, for a token $295,000, it was transformed.

When Jobs began to reassert his grip on the business in autumn 1997, Apple's cash reserves had shrunk to $1.4bn and its debt rating – an indicator of a company's cost of borrowing – had junk status and that October was cut again by the credit ratings agency Standard and Poor's.

Sales slumped by another 28 per cent in the year to September 1997 as its share of the PC market dipped again and price competition bit. The company posted a $1bn annual loss. 'There is no assurance' that ongoing cost-cutting 'will be sufficient to offset the decline in the Company's net sales', Apple warned grimly in its annual report. Rival PC tycoon Michael Dell's advice was to 'shut it down and give the money back to the shareholders'.

Six years on, with its stake in Arm fully dissolved, the company sat on a cash cushion of $4.6bn. Much of that came from a sharply improved operational performance. The launch of new, innovative products meant that Apple was growing again. Healthy sales of 939,000 iPod music players in the year to September 2003 offset a slip in Macintosh sales.

Incidentally, in Arm's first annual report, published in spring 1999, the business dealings of Arm and Apple were laid bare. 'Arm Limited has a contract with Apple Computer Inc., the parent company of Apple Computer (UK) Limited, for the supply to Apple Computer Inc. of software and related support and maintenance to the value of £15,000 in the year ended December 31, 1998,' the report ran.

That sum was flat on a year earlier and sharply down on the £62,000 that Arm had earned from Apple in 1996. And offsetting that modest income was £5,000 of software services travelling in the opposite direction, bought from Apple. There was no indication of the millions of pounds of business the pair would do together in the decades to come – even without ownership to tie them together.

Years later, on 1 June 2010, when Jobs was interviewed on stage at the All Things D conference by technology journalists

Walt Mossberg and Kara Swisher, he revealed the parlous state Apple was in during those first months of his return.

'Apple was about 90 days away from going bankrupt back then,' he confessed, dressed in his trademark black turtleneck and swivelling gently on a red leather armchair. 'It was much worse than I thought when I went back initially.'[14]

This 2010 encounter was bittersweet. It took place days after Apple surpassed Microsoft's stock-market valuation to become the world's largest technology company. But Jobs, peering through round, rimless glasses, was visibly thinner than in previous public appearances and clearly fading. In another 16 months he would lose his battle with pancreatic cancer.

By one measure, Apple had overcome its old foe thanks in part to Microsoft's $150m injection that pulled the firm back from the brink in 1997. But the steady stream of cash provided by Arm, totalling $838m, undoubtedly played a major part in Apple's resuscitation.

Years later, John Sculley, on whose watch the initial invest-ment was made, observed the Arm money 'kept the doors open' for the company.[15] The haul was certainly much more than Apple would have netted if it had accepted Acorn's offer in early 1997. And all from a $2.5m punt on a chip design for a gadget that ultimately failed.

'Book authors like heroes and villains,' wrote Tesler, who died in February 2020, on his website in a critique of another account of this time. 'But we all made a lot of good calls and a lot of ter-rible mistakes during the Newton project.'[16]

It is hard to dispute his conclusion, arrived at in a 2017 inter-view with the Computer History Museum. 'The most successful part of the Newton project was the Arm,' he said, adding that: 'We made more money off the Arm from our stock than we lost on the Newton.'[17]

## Rise and Fall

For Acorn, the impact of Arm's IPO could not have been more different than it was for Apple. Initially, it netted the same £16m from selling down its stake, which Acorn said would be used 'to reduce borrowings and to support the development of its core businesses'.[18] But, as David Lee might have feared, the transaction also left it horribly exposed.

Such was the excitement about Arm, investors had been buying Acorn shares as a proxy so they could get early exposure to the company. Now Arm was listed and its shares were soaring, Acorn's market worth was less than the value of its Arm stake – which spoke volumes about how the City viewed its prospects.

After trying so hard to subsume Arm, Acorn had long since changed tack and tried to persuade it to rescue its former parent on whatever terms it chose. 'We're standing here with our trousers down and there's a tub of butter in the middle of the table,' an Arm executive recalled an Acorn counterpart saying to him during one meeting.

On 28 April 1999, a little over a year on from Arm's IPO, Acorn announced plans to dismantle itself. The company had outlived many of the 1980s PC makers such as Commodore and Atari but after five years of operating losses this was the end.

So that shareholders could exchange their paper for the remaining 24 per cent stake in Arm it held without being hit with an £80m tax bill, Acorn was acquired by the investment bank Morgan Stanley. Its TV set-top box division was sold for £200,000 to rival operator Pace Micro, which also took on Acorn's headquarters building in Cambridge. Apple had earlier bought out Acorn's stake in their Xemplar venture for £3m and the company's RISC operating system was licensed out.

The denouement had necessitated a change of leadership. At a particularly stormy Acorn board meeting, held upstairs at

Cambridge's University Arms Hotel on 4 June 1998, David Lee stood down as chief executive. Notes taken on the day show that Boland labelled the plc board and management team as 'two layers of sludge' that were holding up progress. As the new chief executive, it was Boland who led the break-up. For closing down Acorn's desktop PC division he received letters from loyal but angry developers threatening to beat him up with a baseball bat.

***

The contrast was painful. In their first year of trading, Arm shares more than tripled in value. Spurred on by the proliferation of mobile phones, pretax profits hit £9.4m in 1998 and £18m in 1999 as the number of units sold containing its designs more than tripled to 175 million from 51 million.[19]

In December 1999, Arm entered the FTSE 100, the exclusive club of the UK's 100 most valuable listed companies. Its sky-high £6bn valuation was driven by the technology bubble that would soon start to deflate – but it was easy to see why investors were excited.

A decade earlier, Saxby calculated that Arm needed to shift 200 million units in 2000 if it was to have any chance of becoming the world's embedded RISC standard. Actually, it sold 367 million that year, powered by the mobile boom.[20] Expanding well beyond its initial Nokia alliance, Arm designs featured in two-thirds of all handsets, with more to come. There was a sweet moment in December when Motorola, the unseated mobile leader and Saxby's former employer, bit the bullet and licensed Arm's processor architecture too.

Although almost all the business it won was abroad, suddenly Arm was getting celebrated for what it meant to the UK. On 21 June 2000, Stephen Byers, the trade and industry secretary, opened the company's new headquarters, Arm1, built behind the Cambridge waterworks where the first chip design had been devised 15 years earlier.

Already by that summer, getting on for one-third of Arm's staff, around 150 people, were millionaires. 'Even the administrative secretary is a millionaire several times over,' said Bill Parsons, Arm's human resources director. He estimated that graduates who had been with the firm for three years could be worth £500,000 apiece.[21] In contrast, Steve Furber, the co-creator of the original Arm design, had been given a modest cash amount by the company. Because he was not an employee he could not join the Inland Revenue-approved share option scheme.

Acorn veterans remain stung that their contribution is often airbrushed from the Arm story. After all, it was Acorn that commissioned and incubated the chip architecture, invested several million pounds in it, identified follow-on funding and a high-profile customer, as well as supplying a dozen key staff. And before that, its BBC Micro and Acorn Electron models shifted 1.5 million units, creating a generation of programmers weaned on BBC Basic from which a direct line can be drawn to a UK software industry still vibrant decades later.

———

There was another, last-gasp, reminder of its capabilities. From the ashes of Acorn, Boland raised a small amount of venture-capital money to buy back from Morgan Stanley a team of 30 people that designed components for semiconductors. Renamed Element 14, at its heart was a new microprocessor developed by Sophie Wilson, who had stuck with Acorn throughout. Code-named ALARM – which was short for A Long Arm – the chip could be used to power broadband internet access and was renamed FirePath.

In a final twist, which is remembered differently, Boland had earlier either offered the rump of Acorn to Arm's directors for £1m or pitched the remainder to them as an investment opportunity. Having taken so long to wrestle free from its past, and with some bad blood still evident, Arm declined. A little while later, in October 2000, the US chip giant Broadcom was keener, paying $594m in shares for Element 14.

It was a remarkable outcome, which could only have been bettered if Element 14 had ignored the advice it received from its adviser, Deutsche Bank, and accepted the always acquisitive Intel's equivalent offer in cash. Broadcom stock was soon plummeting – another reminder of the microchip industry's unforgiving rise and fall.

## Chapter 7

# GOING GLOBAL: HOW ASIA MADE THE MODERN MICROCHIP INDUSTRY

## Solid Foundations

In the early hours of 21 September 1999, the people of Taiwan were shaken awake by a 7.6-magnitude earthquake. Striking close to the geographic heart of the East Asian island, devastation unfurled in all directions. Roads buckled, rivers were diverted by landslides, Buddhist temples collapsed and in the capital, Taipei, fires raged, buildings crumbled and hundreds were trapped under rubble.[1]

The tremor was the worst since 1935 to hit Taiwan – also known as the Republic of China (ROC) – and exposed the shoddy construction work of the prior decade's building boom. It claimed 2,400 lives, left 100,000 homeless, cut off power, water and telephones, and bequeathed exhausted residents more than 8,000 aftershocks in the months that followed.

Sixty miles north-west of the epicentre, fate conspired to keep thousands of workers largely safe from harm at the 1,500-acre Hsinchu Science Park, the gleaming home to almost 300 companies and research facilities focused on the high-tech fields of microchips, optoelectronics, telecommunications and biotechnology. At the five factories owned by one of the park's anchor

143

tenants, Taiwan Semiconductor Manufacturing Company (TSMC), the floors rumbled and staff on the night shift were pitched into darkness.

Thanks to a call placed with Taiwan's premier, Vincent Siew, within three days electricity returned.[2] The clean-up from the national tragedy carried on all over the island, with mounting criticism of the government's emergency response. But little more than a week later TSMC's operations were back running at 95 per cent capacity and visitors to Hsinchu – which also benefited from its own independent power supply – could find very little evidence of the recent disruption.

The speed with which it was able to get back to work demonstrated TSMC's great importance to Taiwan. That importance – as a national champion that created skilled jobs, drove exports and the economy at large – was matched only by the key role it had obtained in the global microchip industry. It was why the leaders of major American chip firms such as Qualcomm, Broadcom and Motorola were intently monitoring events from 6,500 miles away.

In a dozen years, TSMC had come from nowhere to produce 5 per cent by value of the global microchip industry's total output. It did not design chips, but instead manufactured them on behalf of some of the biggest names in the business. From its vast fabrication plants, or 'fabs', where silicon wafers were whisked noiselessly from process to process along overhead conveyor belts, came millions of chips to be shipped around the world and installed in mobile phones, games consoles and laptop computers.

TSMC was the first company of its kind, which had contributed to its success, but it was not on its own. Another Taiwanese company, United Microelectronics Corp (UMC), had just sold off its design activities to focus on foundry services and was a close second in size to TSMC. It too was based at Hsinchu, offering another reason for the cluster's political priority on the power grid from which it consumed 500MW per day.

Most microchip firms still made most of the chips they designed in their own foundries, particularly the latest designs. But the sheer cost and complexity of the industry meant the vertically integrated model that dated back to the 1960s was fraying at the edges for older silicon. By proving they could be trusted to closely follow labyrinthine specifications and deliver at scale, foundries for hire had become a seductive option.

It was too late to consider the unwiseness of sending vital work to an island that was prone to natural disaster. Business leaders knew there were alternative geographies to Taiwan, which covered less than one-tenth of the surface area of California and counted 22 million inhabitants, but its ruthless efficiency meant there was rapidly becoming little alternative to TSMC.

---

Less than three months after the quake, the company showed it had not lost momentum. At Hsinchu it held a ground-breaking ceremony for another factory, its first that would be dedicated to handling silicon wafers that were 300mm (12 inches) in diameter.

Utilising these shiny discs would mark a major improvement over the 200mm (8-inch) diameter wafers that had dominated the industry up to this point. Although more difficult and expensive to produce, their increased area – more than twice that of the previous generation – doubled the number of chips that could be produced from each wafer and promised to pave the way for higher-precision designs and higher yields. That would lead to a threefold increase in revenue per wafer and higher profit margins.

Asia's financial crisis of 1997 had sent shockwaves through the region, battering currencies and confidence and bringing to a halt three decades of virtually uninterrupted economic growth. But, by the tail end of 1999, demand for microchips had sprung back thanks to new compact mobile-phone handsets chasing Nokia's success and TV set-top box receivers hitting the market.

The industry had always been characterised by boom and bust, so getting the timing of investment right was critical for commercial success. The 300mm wafer production, which was ideal for cranking out high-volume memory chips, had long been talked about but delayed in practice.

What was different emerging from this downturn was that leading the way with investment were the foundries, not the chipmakers who had traditionally built new plants to give them a competitive advantage. TSMC and peers such as UMC were pushing at the edge of what was possible and, presented with predictions for 30 per cent industry growth in 2000, they were bullish too.

There were signs TSMC was pulling ahead of the pack. In 1999, it had posted another record for both sales and profits despite the loss of one production week to the earthquake. Revenue rose 46 per cent to NT$73.1bn (Taiwan new dollars), or $2.5bn at today's exchange rate, while profits were up 60 per cent to NT$24.6bn, equivalent to $825m. That pleased investors who had bought shares when in October 1997 TSMC had become the first Taiwanese company to list on the New York Stock Exchange.

At the Hsinchu event in December 1999, TSMC's president, F.C. Tseng, said the company was 'dedicated to the pursuit of new technology development and capacity expansion, thereby sustaining our leadership position, increasing barriers to competition, and providing the best dedicated foundry services for our customers worldwide'.[3]

The new facility would add to TSMC's existing estate. It already operated five 200mm fabs and two 150mm fabs. Adding another was not cheap – and nor was payback quick. The cost was more than $2bn – about five times the cost of building a fab a decade earlier – and the new factory was not expected to begin production until early 2002.

The external structure was due to be completed in no more than a year, but then came the technical apparatus. A giant cleanroom and precision equipment would be carefully installed and tested during 2001. At full capacity, the fab would process 25,000 300mm wafers per month, producing many millions of chips. Precisely how many would depend on the chip (also known as die) size, the process node (measured in nanometres) and the complexity of the circuit.

---

As the new millennium dawned, TSMC accelerated its plans. Demand was such that adding 50 per cent to capacity in a year was no longer deemed enough. In the space of a few days in January 2000, the company announced the acquisition of Worldwide Semiconductor Manufacturing, Taiwan's third-largest foundry company, and took control of a joint venture with the electronics firm Acer. Taken together, now TSMC's output was expected to be almost double what it had achieved in 1999 and skittish customers were persuaded the firm was investing sufficiently in the future.

It was a period of great change for Taiwan, as well as the company. In March 2000, 55 years of continuous rule of the island by the Kuomintang (KMT) – also known as the Chinese National Party – was ended when the former mayor of Taipei, Chen Shui-bian, was elected president. His ascent alarmed Beijing because of his support, however carefully expressed, for independence of the island that China had always regarded as its own.

At the end of that month, on 30 March 2000, another cere-mony, this time in the south. TSMC welcomed dignitaries to Tainan, a city best known for its agricultural heritage and strong beef soup, for the opening of another new plant. With Hsinchu running out of space, the Taiwanese government inaugurated the Tainan Science Industrial Park (TSIP) with hopes of closing the north-south economic divide. Marking the occasion with the spectacle of a traditional Chinese dragon dance, TSMC's was the first foundry on site. In case of earthquakes, foundational steel rods had been sunk deep into the bedrock and dampers incorp-orated to offset vibrations.

It was the largest fab in the world. Its cleanroom alone would be 17,600 square metres (190,000 square feet), almost the size of four American football fields. What was noticeable was that as these workspaces expanded, the specifications shrank. Initially the Tainan plant would churn out chips at the 250nm process node, with plans over time to reduce that to 100nm – one-thou-sandth the thickness of a sheet of paper.[4]

To outsiders, the numbers were mind-blowing, but TSMC was certain they added up. The risk associated with such mas-sive capital investment could be spread across a broad range of global customers.

What helped its cause was the broad consensus among equip-ment makers to focus their efforts on the 300mm process over the 200mm process. Faced with a huge investment in manu-facturing if they wanted to make the step up, many chipmakers accelerated the switch to 'fabless', joining a new generation of firms that were springing up and had never had to build their own plants precisely because foundries supplied capacity.

One other, lesser-known, decision from that time was that TSMC declined the offer to license the 0.13-micron – 130nm

– processing technology from the US chip giant IBM. With long-term growth in mind, it preferred to develop its own.

Rather than replicating other firms' best practices as it had been required to do in the early days, TSMC was inventing its own methods to churn out the most detailed, most powerful chips, reliably and cost-effectively. This latest round of investment would extend its leadership position and encourage the further decoupling of design from manufacture.

Asia had long been the workshop of the world, churning out goods for Western consumption. Now that workshop included microchips, which had started out as an emblem of US innovation 40 years ago. But the further operators such as TSMC pulled ahead, the more the US and China were pulled into Taiwan's orbit. Tensions would follow.

## The World Tilts

Within a couple of years of the debut of its Micrologic Flip-Flop at the Institute of Radio Engineers' annual convention in March 1961, Fairchild Semiconductor was preparing to take another step into the unknown.

The defence and space industries provided early sustenance for the microchip industry but there was another market opportunity to pursue: consumer electronics. If Fairchild was going to supply the television and radio market with any success, it had to drive down the price of transistors and scale up volume production.

The company had just opened a plant in Portland, Maine, but Robert Noyce, Fairchild's charismatic leader, had an idea to go further afield. Few of his colleagues knew that outside of his day job he had invested in a small plant in Hong Kong that made transistor radios.

The British colony had long built a reputation as a major centre for the textiles industry and now its consumer electronics output was growing. As the industry expanded, some American firms were experimenting with automating elements of their microchip production line – not just how the chips were made but how the 50mm (two-inch) silicon wafers were moved along the assembly line.

Noyce thought going abroad could be an equally effective strategy. 'He knew the availability of engineers and the wage difference there,' remembered Charles Sporck, who set up and managed the new offshoot.[5]

In 1963 Fairchild took space on Hang Yip Street on the Hong Kong mainland, across the road from a large garment factory. The first offshore assembly plant for the United States microchip industry was hard to miss. Looking out over Kowloon Bay, the large 'F' logo erected on the front of the building could be seen by aeroplane passengers as they flew into the city.

Fairchild made and tested its silicon wafers in the US and then shipped them to Hong Kong, where the chips were assembled, tested again and sold – often going straight back to the US. By the end of the first year, the plant was turning out millions of units to be used in radios. 'It was a winner,' added Sporck. 'I would say that the quality was excellent probably because we had engineers working as foremen.'[6] The internationalisation of the microchip supply chain began with assembly: the fiddly, expensive, labour-intensive 'back end' part of the chipmaking process, that involved packaging individual chips so that they were ready to be sent to customers and embedded in devices. So-called 'bonding' involved attaching tiny wires to each chip that sat in a tiny frame with leads running off it.

This stage was distinct from fabrication – the 'front end' manufacturing of the chip – and, because chips were small, light, high-value items, the two processes soon took place thousands of miles of apart. In a paper for the National Bureau of Economic Research, William Finan of Wharton Econometric Forecasting Associates wrote that in 1973 the manufacturing cost of an integrated circuit was $1.45 per device, if assembly was carried out in Singapore (another former British colony that was establishing itself in the industry). Relocate assembly to the US, and the price soared to $3.

The work wasn't exactly hard. For the most common form of manual tasks, 'assemblers can usually be taught the basic techniques in one day and be reasonably proficient in less than two weeks', Finan added.[7]

It wasn't long before more work went east as Fairchild's Hong Kong facility took on a bigger role within the company.[8] A staff of 6,000 worked round-the-clock on three shifts, blobbing black resin on chips to insulate and protect them from dust and moisture. Customers in the region were dealt with locally and rejected chips were sold to local toy manufacturers for use as teddy bears' eyes, according to Gordon Moore. 'That's probably where the profit came from,' he joked in one interview.[9]

In fact, the economics worked well and Hong Kongers' industriousness was impressive. But its usefulness to the big chip producers didn't last. At first Fairchild paid workers a dollar a day, according to the Stanford University author Leslie Berlin, which was less than their American counterparts earned in an hour. The arrival of numerous Western firms pushed up wages towards $2 a day, blunting the colony's competitiveness just as East Asian locations introduced incentives to attract work. Fairchild opened a plant in South Korea in 1966 offering 80 cents per day.[10]

Chipmakers also experimented with factories in Mexico and El Salvador but appetites fell because of social unrest. Asia, and particularly East Asia, offered greater political stability.

Facilities could run around-the-clock with little disruption. According to one calculation, by 1971 there were 75,000 people employed by the US microchip industry at home but 85,000 in offshore facilities.[11]

––––––––––––

Ironically, the competitive pressure that sent US firms looking for low-cost production centres came from elsewhere in Asia. Japan had been quick off the mark in consumer electronics. As early as 1959, its manufacturers had grabbed 50 per cent of the US market for transistor radios, the first truly portable gadget.

They were promoted heavily. In one US magazine advertisement from Christmas 1960, Sony radios were billed as 'skilfully engineered to produce true, brilliant sound, within a unit of unsurpassed durability and compactness'.[12] One detail that rarely troubled the marketing campaigns was that the major Japanese producers, Hitachi, Mitsubishi and what would become Sony, had licensed US patents to get started.

For a long time, that was the only way US firms could profit from Japan because the country imposed strict import controls. Texas Instruments' hard-won joint venture with Sony, established in 1968, was a rare exception of US–Japan co-operation. It caused a stir because it was the first step towards welcoming foreign investment to the country more broadly.

For the most part, US manufacturers could only peer into this advanced economy where consumer demand for electronics was high. Meanwhile, Japanese firms flooded the US market with their goods, driving down prices thanks to efficient factory techniques.

TI had an entire division devoted to building automated manufacturing kit, such as the Abacus II, which bonded fine gold wire to microscopic contacts on the silicon chip and pin connections on the package. Following the success in sending manual assembly work offshore, the company quickly realised that Japanese technicians could get better results from its equipment too – and at a lower cost.

The cost of the die – in other words, the chip itself, so-called because the wafer was diced to create them – soon overtook labour costs. That was because integrated circuits required more complicated manufacturing than simple transistors, and so automated production lines became essential.

When import controls were lifted in the mid-1970s after sustained pressure from the US, Japan was determined not to lose its industrial advantage. Between 1976 and 1979, its Ministry of International Trade and Industry redoubled efforts to encourage domestic companies to work together. Under the umbrella of the Very Large Scale Integration (VLSI) collaboration, firms pooled research and development. During those years more than a thousand patents were obtained, expertise was fast-tracked and, moving into the 1980s, Japan was set up as a leading supplier of memory chips.

---

Other nations looked on. They saw a fast-moving, high-tech industry still in its infancy and wanted to get in on the action. US purchasing power often worked against its own domestic manufacturers, accelerating the shift east. For example, South Korea had barely made a dent in the market for microwave ovens by 1980 but a US department store came calling, hunting for cheaper goods than it could find elsewhere. When General Electric, struggling to keep pace with the dominant Japanese competition, decided to stop manufacturing microwave ovens and import them under its own label instead, Samsung was the obvious choice.[13]

Encouraged by the government, the South Korean Samsung's electronics division had been established in 1969 to manufacture black-and-white TV sets and refrigerators, one of numerous tentacles of the country's best-known 'chaebol', or family-controlled conglomerate. Lee Byung-chul (B.C. Lee), the son of a wealthy landowner, founded Samsung as a grocery and trucking firm in 1938 but it expanded in all directions, from insurance to retail. He had lived his early life during Japan's occupation of Korea and was intent on building an empire of which the nation could be proud and that would also wean it off imported goods.

Samsung moved into microchips when it acquired a share in the struggling Korea Semiconductor in 1974, a natural fit with its electronics arm that would embrace washing machines, telecoms equipment, video recorders, cameras, disk drives and mobile phones in time.

B.C. Lee had his eyes opened to the challenge in 1982. On a trip to the US to receive his honorary doctorate from Boston University, he took time to tour the semiconductor assembly lines of IBM, General Electric and Hewlett-Packard. It was clear the technological chasm was vast. Lee thought about exiting chips, but ultimately decided to focus on it.[14]

Waking early in his suite in Tokyo's Hotel Okura on 8 February 1983, he called his right-hand man, Hong Jin-ki, and asked him to publish a statement in the Korean newspaper *JoongAng*, another Samsung offshoot.

'Our nation has a large population in a small territory,' he wrote in what became known as his Tokyo Declaration, 'three-fourths of which is covered with mountains but almost completely lacking in natural resources like oil or uranium.' Lee gave thanks for South Korea's educated and hard-working people who had powered the mass export of cheap goods – but warned that trade protectionism meant that that strategy had reached its limit. As a solution, he declared: 'We hope to advance into the semiconductor industry on the strength of our people's great mental fortitude and creativity.'

To test that toughness, 100 Samsung engineers were sent on a 40-mile hike, through day and night, as they prepared a new assault on the market. Rivals were sceptical that Samsung would have much success because a vast amount of investment was needed. In a sign of Lee's determination to move up the value chain, the company would secretly fly chip engineers in from Tokyo on Saturdays to train his team before flying them out again on Sundays.

A pattern emerged across the region. Microchips were aspirational, high-value, highly complicated and a step up from consumer electronics. The four Asian tigers – Taiwan, South Korea, Singapore, Hong Kong – turbocharged industrialisation to catch up with the West. Their appetite to learn was matched only by the taste for cheap imports in the US, whose aid agencies had long sent money back in the other direction.

In the future there was political capital to be had, but for now it was the financial benefits that were obvious. Between 1960 and 1985, real income per capita increased more than four times in Japan and the four tigers. A 1993 World Bank report, 'The East Asian Miracle', put the nations' success and rising living standards down to 'superior accumulation of physical and human capital' and their ability to allocate those to 'highly productive investments and to acquire and master technology'. That stance would continue to pay dividends.

'In this sense there is nothing "miraculous" about the East Asian economies' success,' Lewis Preston, the World Bank president, wrote in the foreword. 'Each has performed these essential functions of growth better than most other economies.'[15]

But Asian progress was not without its tensions. By the early 1980s, the battleground was memory chips. Dynamic random access memory (DRAM) chips that stored in a cell each bit of data or programming code needed by the microprocessor to function were high volume, standardised and required little innovation. They benefited from low-cost, high-yield production, a Japanese speciality.

At first US customers were glad of the second source of the chips when they couldn't match demand themselves. But Japan's supply did not let up when the market slumped in the early 1980s recession. US firms, which were still struggling to sell into Japan, cried foul, alleging that their competitors were pricing below cost in order to grab market share and put them out of business.

It didn't help that for the major Japanese firms, such as Nippon Electric, Hitachi and Fujitsu, chips were only a portion of their sales, and other divisions offered a revenue cushion. For Intel and AMD, chips were virtually all they did. Demands from Motorola for tariffs on Japanese pager and phone imports were granted.

It was a clash over control and security of supply that would echo down the decades. And the US providers were not helped by a ringing endorsement from one of their own. Comments made by Dick Anderson, Hewlett-Packard's vice president of computing, became known as the 'Anderson Bombshell'. At an industry forum in Washington DC in February 1980, he explained how HP had reluctantly turned to Japan when it could not source domestically enough 16K DRAM chips – which stored 16,000 digits of information – for its new computer line. What Anderson quickly realised was that the Japanese chips were more likely to pass HP's factory inspection and the computers they were used in ran for much longer without a memory failure. The imports were far superior.[16]

By the time of the next generation of memory chip, 64K DRAM, Japan had taken 80 per cent of the global market, and for the upcoming 256K DRAM, described by the *New York Times* in 1982 as 'smaller than a postage stamp yet capable of storing all the words on a page of this newspaper', US producers were conceding defeat.[17]

The 1986 Semiconductor Trade Agreement, in which Japan agreed to limit their companies' sales into the US market and make it easier for the US to sell into Japan, brought some relief. But it took 100 per cent tariffs on $300m of goods, including computers and televisions, and firm action by the US President, Ronald Reagan, for the measures to take effect fully.

The move came too late for most of the US DRAM industry. On 11 October 1985, Intel finally threw in the towel, announcing it would withdraw from selling DRAM chips as it posted its first quarterly loss for 14 years. As the last leading memory manufacturer in the US, that left Idaho-based Micron Technology, which had been bankrolled in its early days by John Simplot, a billionaire potato farmer, and his biggest customer, burger chain McDonald's.

Intel had been weighing up its options for more than a year. In Andy Grove's book *Only the Paranoid Survive*, which centres around the concept of the 'strategic inflection point' and the external factors that bear down on a company, he described the moment that he and Gordon Moore, then Intel's chairman and CEO, decided to act. Reasoning that a newly appointed CEO would take the tough decision and pull the company out of memory, Grove said to Moore: 'Why shouldn't you and I walk out the door, come back and do it ourselves?'[18]

———

From being eager to exploit Asian markets, now the US was keen to learn from them. It followed Japan's VLSI project almost a decade later through the creation of the Semiconductor Manufacturing Technology (Sematech), a well-funded, government-backed consortium designed to share research between chipmakers, equipment suppliers and academics.

'You don't want the same thing to happen to the semiconductors that happened to the camera industry and TV industry and so on,' said Charles Sporck, who had gone on to lead National Semiconductor and lobbied the Pentagon and others to explain the significance of microchips. 'Without semiconductors, you're nowheresville.'[19]

Sematech needed a credible, widely respected leader. It called on the very man who had led the first offshoring of work more than 20 years earlier. 'The country had made a commitment to this concept,' Noyce, now leading Intel, said in one interview as he accepted the role. 'I felt that if I didn't follow up with a personal involvement when I was needed, it was betraying a trust to the people of America.'[20]

But the country that the US mobilised against was not an enduring competitor in the industry. Japan's share of the semiconductor market peaked in 1988 once US sales were curbed.[21] 'The Japanese were optimizers not inventors,' said Wally Rhines, who was TI's semiconductor chief during the period. 'As we moved to finding new architectures and new design innovation, they fell behind.' It was the East Asian economies, where government officials did all they could to encourage investment, discourage inflation, devalue currencies, and invest heavily in infrastructure, that became long-term winners and vital trading partners.

Some market entrants also excelled at 'reverse engineering' designs – a euphemism for copying others. It was not until the US passed the Semiconductor Chip Protection Act in 1984 that producing a rival chip with the same design layout as another was made expressly illegal.

***

When Samsung's founder B.C. Lee died of lung cancer in 1987, his company was just beginning to mass-produce DRAM chips capable of storing one million bits – one megabit – of information, only a year and a half after it developed its own 256K DRAM chips. Progress had been exponential since it licensed 64K DRAM technology from the US firm Micron in 1983 after several others turned its overtures down.

Samsung had benefited from state support, including loans that were underwritten by the government. But it was also bold, for example adopting larger, 200mm silicon wafers soon after their instigator, IBM, when Japanese chipmakers hesitated. That meant Samsung took advantage of discounts on new chipmaking equipment that proved to boost productivity. By 1993, having built up its own R&D and skills base, Samsung was the DRAM market leader.

In an echo of efforts to deal with Japanese competition, in 1992 Micron filed an anti-dumping petition on behalf of the US industry claiming that Korean imports of DRAM chips were being sold at less than fair value. Anti-dumping measures were introduced but Samsung was cleared of wrongdoing.

Its rapid path to a leadership position was repeated when an alternative memory technology came to market. Japan's Toshiba developed NAND flash memory – so called for its similarity to the NOT-AND logic gate and because a section of memory cells could be erased quickly, 'in a flash', and rewritten. Fearful it was losing ground to Intel, Toshiba licensed its design to Samsung, which sold its first NAND flash product in 1994 and became the number-one producer in less than a decade.

What began with basic assembly work and some simple technology licensing turned into the market leader in vital components inside little more than a generation. Whether Asian manufacturers 'stole' swathes of an all-American industry, aided by fast following, reverse engineering and slick production, is open to question. The alternative case is that the twin urges of US consumers and Wall Street shareholders encouraged it to be given away.

Either way, global competition blew through the market for microchips and associated electronics. The US could still claim to provide the brains of the industry, but East Asia had applied its manufacturing brawn to give it a vital, unshakeable grip on a complicated supply chain. And one island in particular developed a unique contribution.

## The Least Evil Choice

Within a few months of Jack Kilby joining TI in 1958, another recruit pitched up at its Dallas headquarters whose influence on microchips would prove to be just as long-lasting. Morris Chang, the intense, focused son of a Chinese banker, has less claim to have steered precisely *what* the industry built, but plenty over how and, pertinently, where it did so.

Born outside Shanghai in 1931, Chang moved several times in his youth, including to Hong Kong to escape Japanese bombing during the Second World War. Thanks to some help from an uncle in Boston, in 1949 he arrived in the US to study at Harvard University. Ambitions to be a writer were quickly dropped in favour of engineering, which Chang reasoned was a far more respectable Chinese profession.

He transferred to the Massachusetts Institute of Technology (MIT) but later failed his PhD qualifying exam and instead found work at the small semiconductor division of Sylvania Electric Products, an established name in electronics that had mass-produced vacuum tubes in wartime. There, Chang's task was to improve the reliability of germanium transistors rolling off the production line. Having studied William Shockley's 'Electrons and Holes in Semiconductors with Applications to

Transistor Electronics', he later designed transistors himself. But frustration at Sylvania's marketing efforts caused Chang to quit for TI after three years.

In Dallas, he was soon propelled into a management role after boosting the output of a troubled production line that was dedicated to transistors bound for IBM, a key customer. By tweaking the temperature and pressure at various points of the process, the yield went from virtually zero to 25 per cent. 'Suddenly we were in business, whereas before, we were just cranking out rejects,' Chang said.[22]

Some of Chang's innovations, such as slashing the price of new products early on to win market share, became commonplace across the industry. They propelled him up through the organisation to run its semiconductor division, two levels below the CEO, and he garnered a fierce reputation for detail so that his plant inspections were akin to royal visits.

Chang's career trajectory fomented dreams of one day becoming CEO himself, even though TI was unlikely to choose a non-US citizen to run a strategic national defence supplier. But it was the group's plan to move determinedly into consumer electronics in the mid-1970s that ultimately derailed his ambition. Even though he was installed as head of the consumer division, which made calculators, watches and home computers, Chang felt miscast.

'The customer set – completely different. The market – completely different,' he said. 'And what you need to get ahead in that business is different, too. In the semiconductor business, it's just technology and costs; in consumer, technology helps, but it's also the appeal to consumers, which is a nebulous thing.'[23]

Still, on his watch, Speak & Spell, the speaking electronic toy that used the world's first single-chip speech synthesiser, was a hit – but the division declined. Feeling sidelined, Chang quit in 1983.

A senior role at another chipmaker, General Instrument in New York, had not fulfilled its potential and Chang had separated from his wife, Christine. So when the call came from Kwoh-Ting (K.T.) Li, a Taiwanese government official, he was at a loose end. After all these years in the US, the offer to go back east to run Taiwan's Industrial Technology Research Institute (ITRI) was about as leftfield as it got.

Chang knew Li, having set up an assembly and test plant in Taiwan for TI. He was something of a visionary who thought ITRI could become as celebrated for its cutting-edge research as Bell Labs was in the US. The plan was typical for Taiwan's former finance minister, who was fond of the big picture. Li had studied radioactive substances under Ernest Rutherford at Cambridge University and later focused his work in the area of superconductivity. But his physics career had segued into politics and economics by the time he fled from the Chinese Communist Party to Taiwan in 1948.

After a short spell as a Dutch colony, Taiwan had been administered by China's Qing dynasty for almost two centuries, switching to Japanese control in 1895 at the culmination of the First Sino-Japanese War. When the second Sino-Japanese War ended in 1945, Japan surrendered Taiwan to China. Its sovereignty was further complicated when China's ruling KMT party retreated to the island in the wake of its defeat in the Chinese Civil War in 1949. Martial law was imposed and KMT began ruling with the consent of its US and UK allies, but China was clear that Taiwan had broken away and would someday become part of the country once again.

As it fought to forge its own identity, Li felt strongly he could use science to benefit Taiwan. 'The war had a profound impact on my life, but I never regretted giving up my academic research,' he said.[24] Just like Samsung's efforts in South Korea, here was something to help Taiwan define its place on the world stage.

It had become known for much more than agriculture during the 1970s. Drawn by the low cost of labour, General Instrument had been one of the first to set up a factory there that built TV components, but not semiconductors. However, the opening of the Chinese economy to the outside world in 1978 saw Taiwan's appeal to inward investors fade. In the 1980s, the appreciating currency and rising wage costs meant some firms began moving production away in search of a better deal.

Taiwan was intent on shifting up a gear, from making computers and monitors to the key parts they contained. Since 1976, Li had been the minister without portfolio with a remit to expand Taiwan's high-tech industries. ITRI had already licensed some semiconductor processing technology from RCA Corporation, once the Radio Corporation of America, which had for decades been the dominant consumer electronics firm in the US but was now feeling the heat from foreign competition.

In addition to selling a licence, RCA had agreed to train Taiwan's engineers, which was described in one ITRI account as 'the decisive moment in the semiconductor industry's 30-year history'. Technology transfer, through reassigning rights, cooperation or investment, was one thing, but training was 'less direct and subtle'. It had a 'nonsaleable quality', the report said. 'In other words, the most valuable outcome was the "hidden" knowledge these ITRI trainees brought back.'[25]

Li had more in mind: providing tax incentives to firms that invested heavily in research and development, strengthening science, engineering and computing education and recruiting technical manpower from abroad. These included former residents that had gone overseas – and friendly industry leaders such as Chang.

Li also consulted Stanford University's Frederick Terman on how Taiwan could emulate the success he had overseen in Silicon Valley. The result, opened in 1980, was the Hsinchu

Science-Based Industrial Park, to which companies were lured to the first 500 acres with tax holidays, cheap loans and discounts on land. Hsinchu was 'a symbol of the determination and confidence based on careful planning', the *Japan Times* reported, adding that Taiwan's goal was to upgrade 'its industrial and trade structures to the level of advanced industrial countries like Japan and the United States'.[26] It was located close to the National Tsing Hua University and National Chiao Tung University, both vital sources of new recruits.

In luring Chang, the premier of Taiwan, Sun Yun-suan, also an engineer by training, told him that he 'particularly wanted to use my ability to transfer technology from just research results to economic benefits for Taiwan industry'.[27] Sure enough, within months of his arrival, Li upped the ante. The businessman was already moving ITRI towards sourcing some of its funding from corporate contracts in place of pure government subsidy. Staff who thought they had jobs for life discovered that was no longer the case.

That was all very well, but the government didn't just want a commercially focused research centre, it wanted to develop its own semiconductor company, Li explained. At a few days' notice, Chang was invited to prepare a business plan and tell them how much money he needed.

---

He surveyed the market and assessed Taiwan's strengths and weaknesses. Chang knew from his time at TI that this industry's pace of change was breathless. If Taiwan wanted to progress it would have to battle with long-established industry giants including TI, Intel and AMD. Yet the island's firms were hopelessly behind in research and development, circuit design, and sales and marketing techniques. And no one was waiting for them to catch up. Take the RCA technology: it was behind the industry frontrunners when ITRI licensed it in 1975. A decade on, and without proper competitive pressure, it lagged even further behind.

Taiwan's only possible strength was in manufacturing, which Chang regarded as 'the least evil choice'.[28] The electronics industry had located their assembly and manufacturing work in Taiwan for years. It made sense to fashion a company that could independently manufacture for the chipmakers.

Chang had another thought. During his time at TI and General Instrument, he knew of many chip designers who would have broken away to start up their own venture if it hadn't become so expensive to do so. They either had to invest heavily in their own manufacturing facility or wait in line for excess capacity on the major firms' production line. Some drove around Silicon Valley with a box full of wafers looking for those people that ran factories. When they found them, the price demanded often involved sharing their hard-won intellectual property. Perhaps Chang could offer them an alternative.

In Taiwan in 1987, change was in the air. Direct military control was lifted by the KMT that July after 38 years as relations with China showed signs of improvement. Five months earlier, when the Taiwan Semiconductor Manufacturing Company opened for business in February 1987, it had attracted significant backing. Chang had raised $220m: half from the government and more than half of the rest from the Dutch chipmaker Philips, which also licensed some of its technology to the venture. Chang's status as a foreign expert was key to bringing on board local investors for the remainder.

There was scepticism over whether the model would work. Intel, TI and several Japanese semiconductor companies declined to invest, Chang recalled years later in a Stanford University talk.[29] The received wisdom was that chip design and process technology needed to develop hand in hand for chipmakers to succeed. Splitting them just complicated matters, particularly when intellectual property was jealously guarded.

'Real men have fabs,' declared Jerry Sanders, the flamboyant chairman of AMD, who firmly believed in the importance of owning manufacturing. The famous comment is also ascribed to T.J. Rodgers, another outspoken industry leader and boss of Cypress Semiconductor, a firm based in San Jose, California.

At first TSMC survived on crumbs from the biggest chipmakers, including Intel, Motorola and TI, who would gladly pass on the manufacture of their oldest technologies that still enjoyed residual demand, if it freed up capacity in their own facilities to produce the latest designs. They were not disappointed: '. . . everybody expected . . . this new company to be a weak competitor,' Chang said years later. 'Well, they found this new company, TSMC, to be a strong supplier.'[30]

From the early 1990s, the fabless start-ups designing specialist chips began to bring in business. They were keen to work with a manufacturer that was not in competition with them. For these newcomers, including California-based Nvidia, established in 1993 among a crop of specialist graphics chipmakers, TSMC's arrival was transformational. The opportunity to invest faster in R&D accelerated years of progress for the industry – and ultimately for the end-user. Chang estimated that the community of 25 fabless companies that existed as TSMC set up had expanded 20-fold a decade later.[31]

TSMC won copious business but did not keep it without ferocious attention to detail. One Friday afternoon in 1999, Nvidia's co-founder Jensen Huang received a visit from Chang, unaccompanied, carrying just a pencil and black notebook. He was in the area and checking up on a valued customer. Huang only discovered later that Chang was in the US on honeymoon with his second wife Sophie.

The success of TSMC realised Li's ambitions to slow down the brain drain. During the 1970s and 1980s, thousands of Taiwanese students went to the US, pulled by fellowship money to pay for graduate studies and pushed by the limited career opportunities at home.

According to research by the National Youth Commission, by 1998 more than 30 per cent of the engineers who studied in the US returned to Taiwan, compared to only 10 per cent in the 1970s. One of those to return in 1993 was Mark Liu, who had been researching undersea communications cables at Bell Labs in New Jersey. His first challenge at TSMC was to build a $1bn factory; by 2013 he was co-CEO and later chairman. The island had become a magnet: for workers, for customers, for those envious to learn.[32]

Chang's light-bulb moment that TSMC could not hope to do everything in a costly and fast-moving market led it to achieve operational excellence in just one vital area. Supported by the Taiwanese government's 20-year focus on building its global position, that shattered the structure of the microchip industry, as did Chang's offer to work with anyone.

TSMC was an enabler, not a competitor, unleashing new chip-making talent and reviving long-established firms by handling production for them. It did not seek to exert control or political influence, which only accelerated its progress.

In its own way, Arm was fast becoming another firm in a similar mould.

# PART TWO

## ARM (2001–16)

## Chapter 8

# INTEL INSIDE: A PC POWERHOUSE TRIES TO MOBILISE

## Goodbye, Grove

The hundreds of shareholders that filed into the Santa Clara Convention Center on 18 May 2005 knew they were about to witness a little piece of history. Many had reprised this pilgrimage year after year, often choosing the same part of the auditorium to sit in and gaze up at the illuminated line of speakers. Many in attendance were former employees who owed most of their personal wealth to this famous American company. Many had made a vital contribution to build it.

One man had contributed more than most in attendance, and it was his farewell from Intel that would dominate the microchip giant's stockholders' meeting. Andy Grove, the last of the legendary trio to have steered the firm from its inception in 1968, was taking his leave as chairman after eight years, having also served 11 years as all-conquering chief executive. In all, he had spent 37 years with the company.

'I've seen Andy as provocateur par excellence, visionary, charismatic leader, technologist, teacher, father, grandfather, author, policy advocate, philanthropist,' said David Yoffie, a Harvard Business School professor and Intel's lead independent director, who led the flow of tributes that day. 'Perhaps more important, Andy has never been satisfied with the status quo. He endlessly pushed himself, and everyone around him to change the world.'[1]

Craig Barrett, a 31-year Intel veteran who was stepping up from chief executive to chairman to succeed Grove, contributed to the congratulatory mood, adding: 'I consider him to be my mentor, and . . . the image of what I think a great manager, and a great person, a great human being can be.'[2]

Even the somewhat cantankerous Grove, with tousled grey hair, rimless spectacles and an impassioned delivery flavoured by the strong accent of his Hungarian homeland that he fled as a penniless refugee, was moved to share some warm words as his long-standing colleague Gordon Moore, now Intel's chairman emeritus, looked on. The admired business leader and industry sage was credited with the risky, mid-1980s gear change from memory chips to microprocessors, swapping a saturated market for a nascent one, that put Intel on the path to great success.

Grove paid tribute to 'particularly our employees, present and ex-employees, who have made it fun 80 per cent of the time and who have made me proud of being part of Intel'. Turning to his family, he thanked 'my wife Eva, my daughters Karen and Robbie [who] accepted the presence of Intel as a sibling, or as a mistress, more or less good-naturedly'.[3] They, together with a handful of close colleagues, had also helped him come to terms with the diagnosis of Parkinson's disease that would not become public for another year.

It was easy to be awed by what Grove was effectively leaving behind at the company that day (although he would take the title of Intel's senior adviser). With a vast headquarters that lay a five-minute drive from the conference centre down Santa Clara's Great America Parkway, Intel was a money machine. In 2004 it posted record revenue of $34.2bn, net income of $7.5bn – a one-third rise on the previous year – and paid out a best-ever cash dividend of $1bn.[4] Robert Noyce, the prime mover in Intel's early days who had died in June 1990, could never have dreamed the company that his old partners Moore and Grove now surveyed would scale such heights.

Nor was Intel resting on its laurels. The company had sunk an astonishing $4.8bn into research and development work in the preceding 12 months. Never mind that most of its microprocessors were manufactured using 90nm process technology – featuring structures smaller than the size of a virus, the world's smallest micro-organism – Intel was ploughing cash into its 65nm technique whose production was due to be ramped up in the first half of 2005. With great pride, it declared in the annual report that year that its researchers believed they could continue extending the homegrown Moore's Law 'for at least another 10 to 15 years'.[5]

Solidity, continuity and great financial firepower: Intel appeared to be as dominant as it had ever been in the chip industry that it had done much to shape. And yet as the applause rang around the hall – in a rare moment of celebration for the no-frills firm revered for its hard-driving culture – trouble was already brewing for the men that Grove left in charge.

## A New Rivalry

Pepsi and Coca-Cola, Amazon and Walmart, Adidas and Nike: there are some legendary business wars that have enlivened the corporate world for decades. They deplete returns for shareholders, spark innovation for consumers, flash dollar signs in the eyes of advertising executives and generate colourful copy for financial journalists. The best rivalries come steeped in admiration; they motivate each side to up its game.

Intel already had a nemesis. Its long-running battle with Advanced Micro Devices (AMD) was a fraternal scrap that belonged to a plotline from *Dallas*, the frothy Texas oil saga, except it ran for years longer than principal characters J.R. Ewing, his brother Bobby and archrival Cliff Barnes could sustain their own hostilities.

Any other rival would have been closed down by the litigious Intel if it was found to be reverse engineering – in other words, copying – its in-demand 8086 chip design. But AMD was led for many years by Jerry Sanders, a flamboyant former Fairchild Semiconductor salesman known for his fancy cars and crisp linen suits who looked upon Robert Noyce as a father figure long after he had departed to co-found Intel. The affection was reciprocated to the extent that in 1976 Noyce suggested down-on-its-luck AMD should be granted a licence to become a 'second source' manufacturer for the 8086 chip, something that large customers were increasingly demanding from their chip suppliers as an insurance policy to guard against shortages.

'We know them, they don't have the manufacturing capacity to challenge our dominance,' was Noyce's thinking, according to the author Michael Malone, '. . . and we can handle Jerry. After all, he's family.'[6] This extraordinary act of munificence granted AMD a business model and Intel decades of legal wrangling.

Grove did not share Noyce's affection for Sanders. When, in August 1991, AMD launched a $2bn lawsuit that alleged Intel had acted unlawfully to monopolise the market for PC microprocessors by threatening to withhold its more advanced chips from computer makers who bought from its closest rival, Grove decried the firm as 'the Milli Vanilli of semiconductors', invoking the lip-syncing, Grammy-winning pop duo, adding: 'Their last original idea was to copy Intel.'[7]

───────────

Rather than having any fraternal connection, Arm and Intel had started out from very different places, serving different markets. But as mobile telephony merged with personal computing they were drawn inexorably into each other's orbit.

Intel was Chipzilla, so nicknamed as the industry giant that smaller competitors loved to hate for its desire to dictate how the industry developed. It jealously guarded its intellectual property and thought nothing of dragging its opponents to court.

Arm was consensual and quirky, a rare British interloper in a sector driven by all-American ideas and increasingly slick Asian manufacturing. It could be protective too, but its mantra was to be open and serve everyone if they paid for a licence and device royalties. In fact, Intel and Arm didn't directly compete: Arm merely licensed its processor designs to Intel's competitors and, eventually, to Intel too. But its mobile prowess deprived the American giant of another market to dominate that would prove to be far larger than personal computing alone.

---

What coloured their relationship further was that long before it became a competitive threat, Intel tried to buy into Arm. One day in 1994, Robin Saxby picked up the phone to discover Les Vadász, Andy Grove's right-hand man at Intel, on the other end of the line. Even though there had been no warning of the conversation, Vadász, who ran Intel Capital, the microchip giant's free-spending venture arm, came straight to the point.

'I hear you want to sell Arm,' Saxby recalled hearing in the unmistakable Hungarian accent Vadász shared with his countryman Grove. It was the kind of call the Intel veteran had made to numerous promising start-ups since its investment activities had been formalised a few years earlier.

To keep track of the newest technological developments and fortify its own market position, Intel regularly took stakes in smaller firms, particularly those active in computing, networking and the internet, which it saw as growth areas for its microchips. Citrix, Broadcom and Cnet had been notable hits, but one famous miss was Netscape, an early web browser that had prospered until Microsoft swamped the market with its own product.

On this occasion, Vadász's opening gambit could not have been further from the truth. After a sticky start, Arm was gaining traction with customers, generating cash, and the Nokia work via TI that would prove to be a gamechanger was progressing well. Three years into his leadership, Saxby could already envisage a stock-market flotation that would reward staff for their hard work and loosen ties with Arm's founding shareholders.

Acorn had other ideas, as did Olivetti, Acorn's major shareholder, which looked to be a willing seller of Acorn, its key Arm asset, or both if they solved its own financial woes. Hence Intel's curiosity.

Saxby admired Intel for its relentlessness and how it had dominated the PC market with its x86 family of chips. His ambition was to do the same – albeit using very different methods – by establishing a global standard for mobile and other low-powered devices. 'Britain hasn't yet had an Intel-style success,' he said in an interview that same year. 'I think it's about time.'[8]

Of course, Arm would likely never have been born if Intel had agreed to license the 80286 design to Hermann Hauser for the Acorn Archimedes. But with the wind in its sails, this was no time to surrender its independence to the larger firm. Actually, despite picking up momentum across several markets, Intel was one chipmaker not yet working with Arm designs.

'Les, we're doing really well,' Saxby said in response to Intel's pointed inquiry. 'We don't need you to buy us and you don't need to buy us either. You should just take a licence.'

The response was enough to cool Vadász's immediate interest. 'I can see the Arm plane is going fast down the runway and you don't want us to stop it taking off,' Saxby recalled him answering.

## Otellini's Challenge

Arm versus Intel was thrown into sharp relief during Paul Otellini's years at the helm of Intel.

On 23 August 2005, three months after his coronation at the stockholders' meeting where investors bade Grove farewell, Otellini made his first big address as chief executive at the Intel Developer Forum (IDF), a biannual event designed to excite partners, customers and Wall Street investors about upcoming Intel products.

Striding on stage in San Francisco, he began by reminding the audience what he had said four years earlier at the IDF. 'I talked about the notion of moving beyond gigahertz, the classic focus of Intel,' Otellini said. Now he was following through. Rather than simply seeking to boost the power of its chips, generation after generation, 'we need to think about delivering performance against a new metric, and that is performance per watt'.[9] The power consumption of its chips would fall by a factor of ten by the end of the decade, Otellini pledged, while at the same time performance would increase tenfold.

Even though he was umbilically linked to Intel's foundations, the announcement of his leadership appointment in November 2004 had been viewed as a break from the past. The new chief executive was no newcomer, joining the business in 1974, the same year as Craig Barrett and just as a giant round of layoffs had been announced. His selection in 1989 to serve as Andy Grove's technical assistant and de facto chief of staff put him on a trajectory to the top. But the reserved, bespectacled Otellini was not an engineer like his four predecessors. His degrees were in economics and business and he started out as an analyst in the finance department, later managing Intel's key relationship with IBM, running sales and marketing and then the microprocessor business.

His San Francisco announcement was an acknowledgement that the market was moving. The focus on power, which had made Intel so powerful over the last 40 years, was no longer enough. The mobile revolution had untethered devices from the wall socket. Using gadgets on the go meant that power efficiency was of prime importance for customers and consumers now. It was no use stuffing hardware with high-performing chips if they were going to drain energy in a hurry.

Searching for chips that had a longer battery life, Otellini thought he had found the answer in Israel. At Intel's facility in Haifa, the country's third-largest city, developers far removed from the power-hungry traditions of Silicon Valley had come up with Centrino, a family of sleeker, simpler chips that gave off far less waste heat than the company's existing desktop range. Since their launch in 2003, the proportion of Intel's PC microprocessor sales that were tailored specifically for mobile devices had risen from 23 per cent and were forecast to hit 36 per cent in 2005.

The notebook computer – a lightweight, portable laptop – was gobbling market share and had in the previous quarter for the first time outsold the traditional desktop PC that had been the cornerstone of Intel's vast profits for a generation. Otellini was hopeful that Centrino would be the company's salvation as the gadgetry developed further towards what he labelled 'hand tops', which would combine 'the performance capability of the PC with the mobility that you get in a handset today'.[10]

In truth, Centrino was just about all that had paid off for Intel recently. The company had been struck by product delays and manufacturing problems. By avoiding layoffs during the latest industry depression, Barrett had just left them for his successor to deal with as he stepped up to become chairman. And as well as bequeathing him a bloated payroll, Barrett had also splurged $10bn on communications businesses designed to accelerate Intel's mobile effort, but they remained resolutely loss-making while Texas Instruments made hay with Nokia. Even Advanced Micro Devices (AMD), Intel's long-running rival, had struck gold in 2003 with the Opteron chip that had the effect of slashing prices in the high-end market for computer servers.

The key problem was that the company famous for enabling technological change was not changing fast enough itself. These days chipmakers and device manufacturers were collaborating on designs, rather than one dictating to the other as had been the case in Intel's pomp. And loath to sacrifice fat margins, the company had to work out not only how to make low-powered, low-cost chips, but also how to make a profit from them when the device could sometimes sell for less than it cost to build a high-end processor.

'How do we fit inside of something that sells for $100 and make some money?' Otellini pondered in a 2007 interview.[11] He might just as easily have asked: How does Intel become more like Arm?

## The Power of Wintel

Before detailing how Intel and Arm butted heads it is worth examining where Intel's sustained success stemmed from as one half of perhaps the most successful corporate partnership of all time. For that, it has Operation Crush to thank. The inspired and aggressively named programme was the response to early signs that Motorola's 68000 microchip – the same type that Robin Saxby would fail to sell to Acorn's Hermann Hauser and Chris Curry on their trip to East Kilbride – was performing far better in the market than Intel's own 8086 16-bit processor that had been launched in 1978.

This discovery was made all the way back in 1978. In the eight years since Ted Hoff and Federico Faggin had created the 4004 microchip, Intel had progressed through several cycles of Moore's Law. The 4004 featured 2,300 transistors and was built using the 10 micrometres process, which indicated the space – one-hundredth of a millimetre – between each of them. The 8086 was packed with more than 10 times the transistors and 4,000 times the maximum memory capacity.

Processors had found a home in calculators, games consoles and hobbyists' build-your-own computers, and with every improvement in pricing and performance new market opportunities emerged. So did competitors. Not only did Intel lack another product to meet Motorola's threat, but it was also concerned about chips being offered by Zilog, a rival firm Faggin had quit to co-found in 1974.

It turned to its marketing experts for an answer, and dangled the promise of exotic trips to Tahiti for its salespeople that delivered orders. Without each rep being pushed to sign one new customer every month, one enterprising agent would never have thought to approach the unlikeliest of customers, a leading force in the developing computing industry with its own vast, in-house chipmaking capability: International Business Machines (IBM).

---

The salesman, Earl Whetstone, called on the company's facility in Boca Raton, Florida, 'pursuing a rumour that Big Blue was working on a black-box project that might involve microprocessors', the technology author Michael Malone wrote in his account of Operation Crush.[12] Whetstone got lucky. After a couple of years sitting on the sidelines, IBM – whose nickname stemmed from the colour of its mainframes, user manuals and corporate logo – was now taking the personal computing market seriously enough to consider entering it. 'How do you make an elephant tap dance?' the group's chief executive, Frank Cary, famously said. Aware it had to move fast to catch up with nimble competitors, a new unit outside IBM's normally slow-moving product development process was set up.

Under the project leadership of Don Estridge, an IBM engineering manager, a cheaper version of Intel's 8086 chip, the 8088, was chosen for the new device, which was ironically codenamed Acorn, presumably with no knowledge of a certain British company starting out in the same field.[13] Alongside Intel's chips there was an operating system called MS-DOS, acquired and adapted in a hurry by Microsoft, at the time a firm with just a few dozen employees led by a geeky 24-year-old, Bill Gates.

It was a notable shift in product. IBM had enjoyed great success during the 1970s selling mainframe computers that cost up to $9m, required a quarter of an acre of air-conditioned space and 60 people to run and keep them loaded with instructions.[14] Where technology was concerned, the company was a comfort blanket for corporate executives scratching their heads at these new portable devices, hence the unattributed expression, 'Nobody ever got fired for buying IBM.' Now the company had in mind a device that straddled work and home, in comparison to its cabinets a tiny machine costing $1,600 that carried out business tasks but also plugged into the TV for video game-playing.

No one could have predicted the impact the IBM 5150 personal computer, known simply as the IBM PC, would have when it was unveiled on 12 August 1981 at New York's Waldorf-Astoria. Also worth noting was that this latest launch came seven months after the US Department of Justice had dropped its effort to break up IBM – an investigation that may well have shaped its newest product. Without regulators circling, it is easy to doubt that IBM, which had designed and built every computer it had sold up to this point, would have gone about things the same way.

No matter. IBM's original forecast that it would need to produce one million machines in the first three years, with 200,000 in the first 12 months, was a significant underestimation. Its reputation for business machines meant the device was an easy sell for corporate clients and its trusted brand broadened the market for PCs almost overnight. Apple, Acorn and the crop of pretenders that had focused on the home and education markets were eventually swept aside. Customers were buying 200,000 PCs per month by the second year and IBM soon grabbed 80 per cent of a growing market – making it at least as dominant as it had been in business machines.[15]

What differed this time was IBM's willingness to help rivals along. It couldn't be accused of cornering the market if it offered up its source code to encourage the manufacture of copycat PCs. These 'IBM compatibles', based on the published circuit designs and source codes, helped to create scale and a new industry standard for which applications including games were enthusiastically developed. So successful was its strategy that *Time* magazine eschewed its Man of the Year title for 1982, instead crowning the IBM PC 'Machine of the Year'. The publication declared that: 'the enduring American love affairs with the automobile and the television set are now being transformed into a giddy passion for the personal computer'.[16]

However, IBM's stroke of genius did not endure. The price of the so-called clones made by Dell and Hewlett-Packard kept falling and before long Big Blue was struggling to keep pace with the market it had lit the touchpaper on. Rivals such as Compaq adopted upgraded Intel chips faster than IBM chose to. That decision was curious because the performance improvement was stark. The 80386 microprocessor was the first 32-bit processor produced by Intel and its 275,000 transistors were nearly double the amount the 80286 featured.[17]

---

It came to pass that even the biggest company in the world, accustomed to powering the Dow Jones Industrial Average to fresh highs during the 1980s, could lose out in a cut-throat industry. Selling computer 'boxes' became a commodity that was difficult to profit from. In December 2004, with market leadership a distant memory, IBM offloaded its PC business to Lenovo of China. In the same week, *Computer Almanac* calculated that IBM-compatible PC sales had reached $3.1tn overall, based on 1.5 billion computers sold.[18] It was an astonishing number. IBM's first chairman, Thomas Watson, is reported to have predicted decades earlier, 'I think there is a world market for maybe five computers,' although attribution of his comment is hard to trace.

Long before the Lenovo deal, it was clear who the long-term winners of the 1981 launch had been. IBM allowed Microsoft to retain the rights to its operating system and licence to other PC makers that crowded into the market. Nor were restrictions placed on Intel. In fact, IBM's support went a step further for its chip supplier. In late 1982, it paid an above-market $250m to take a 12 per cent stake in Intel, providing vital cash to see it through a period of weak demand for memory chips and until sales to PC makers took off. The stake rose to 20 per cent but was sold down five years later.

The industry clustered around what became the x86 chip family, so it became a common computing standard and its speed, capacity and the relationship between price and performance made exponential progress. It happened not just because Intel championed and set about improving the x86 line – it was because hundreds of other companies vied to do the same thing. Together with Microsoft Windows, Intel was installed and licensed across generation after generation of devices. Yoked together, the pair earned billions of dollars every year, even once the catalyst of IBM had faded. Better still, PCs gained fresh impetus when the internet came along.

It wasn't always a cordial relationship. Microsoft's chief executive Bill Gates and Andy Grove clashed over Intel's ambition to drive deeper into software. Gates wrote to Grove at one stage, saying that 'when Intel finds someone who has some humility about developing operating systems and the complexities involved then maybe we can try to work together'.[19]

'Wintel' – a portmanteau nickname neither party liked – reached its apotheosis in August 1995 when Microsoft debuted its Windows 95 operating system. Described by the *New York Times* as 'the splashiest, most frenzied, most expensive introduction of a computer product in the industry's history',[20] it was the moment personal computing crossed defiantly into the mainstream.

New York's Empire State Building was lit in Windows 95's colours of red, green, blue and yellow. Microsoft paid several million dollars to use the Rolling Stones' song 'Start Me Up' as the soundtrack for the system that featured a 'start' button to organise desktop applications into groups. The US chat-show host Jay Leno, with bouffant grey hair and baggy chinos of the era, presented a glitzy launch event in Redmond, Washington. Microsoft executives led by the bombastic executive vice president Steve Ballmer broke into an awkward on-stage dance routine as the music blared. Consumers were just as excited. In the first year, they bought a record-breaking 40 million copies of the software that synced exclusively with Intel hardware.

Intel's leaders did not need to take to the stage for the company to develop its own swagger. Once a technological curiosity for the few, the power of its intricate circuitry was emblazoned across newspapers and TV commercials long before Windows 95 grabbed the media's attention. The 'Intel Inside' advertising campaign debuted in July 1991 with a five-note audio jingle that became instantly recognisable for years to come.

It swiftly became a kitemark of performance and quality, a simple reassurance for consumers who really knew nothing about the complicated innards of the device they were considering buying. If the only way for the market leader to carry on growing was to grow the market, then some advertising buzz was bound to stimulate interest and influence purchasing decisions from downstream.

Just as IBM's openness had created a broader PC market, Intel was clever in how it used its direct customers – the computer manufacturers – to spread the word of its own effectiveness. In return for including its blue-circle logo in print advertisements and stickered on the side of its PCs, makers earned back 5 per cent of the purchase price of Intel's processors. Those funds were diverted into a pool to jointly pay for advertising. By the end of 1992, over five hundred manufacturers had signed up to this co-operative marketing programme. Some 70 per cent of manufacturer ads that could carry the logo did so.[21]

Not everyone liked the idea of promoting so prominently a supplier that might dent their own brand value, however. Some grumbled they had little choice but to sign up given Intel's power in the marketplace. IBM and Compaq pulled out of the programme at one stage, with Compaq pointedly stating on the cover of its 1995 annual report: 'When it says Compaq on the outside, you don't need to worry about what's on the inside.'[22]

How ironic that Intel was accused of being a corporate bully with a campaign designed to popularise its technology but with the knock-on effect of lightening its image. Nonetheless, it had the desired effect. Despite some scepticism about the strategy of 'ingredient branding' and an initial $250m outlay, Intel's share of the PC microprocessor market rose from 56 per cent to 86 per cent by 2001.[23]

As it swept aside allcomers, Intel grew bolder still. The brand was splashed on billboards and thousands of bicycle reflectors distributed in China. During the Super Bowl on 26 January 1997, when the Green Bay Packers ran out winners against the New England Patriots, TV viewers were introduced in the interval to the Intel dancers, a group of gyrating scientists in bunny suits. To the disco soundtrack of Wild Cherry's 'Play That Funky Music', the commercial claimed of Intel's new Pentium MMX processor: 'It'll make your multimedia dance'.

## StrongARMed

Intel's lawyers were kept nearly as busy as its marketers. There was the ongoing battle with AMD, plus the Federal Trade Commission (FTC) that in March 1999 set out its antitrust case against the chipmaker that alleged bullying of three of its customers to maintain a stranglehold on the market.

One of that trio, Digital Equipment Corporation (DEC), had an illustrious past making VAX minicomputers that were once beloved by scientists and engineers who did not need an expensive mainframe to carry out their tasks. Now it was working closely with Arm to try to recapture former glories.

Spotting Apple's launch of the Newton MessagePad in 1993, DEC's engineers turned their attention to developing something more efficient for the PDA market that might in time interest Apple or its competitors. The low-power version of its own Alpha chip 'turned out to be an interesting technical idea but not an interesting marketing and business idea', said Dan Dobberpuhl, one of DEC's senior engineers that led the project. 'So that concept kind of died and we all concentrated on the idea of building a high-performance version of an existing low-power chip. We ended up, for various reasons, choosing the Arm architecture.'[24]

Within a couple of years, that project become a promising joint venture. When DEC shared its early work with Arm, both sides realised it could be mutually beneficial. DEC extended Arm's first 'architecture licence', which permitted significant variation from its designs while maintaining compatibility with its instruction set, and the two sides vowed to work together to design and build new microprocessors. 'I said to the board, we have to grasp this,' said Dave Jaggar, the architect of the Thumb extension. He knew some of the engineers involved and observed almost half of his Arm colleagues were still tied up developing a final chip for Acorn instead of pressing forward into new areas. 'These guys were the sort of microprocessor architects we were not: well experienced, a really tight-knit team.'

A press release issued on 21 February 1995 declared that a new family of 32-bit RISC products, christened StrongARM, was 'intended to complement and broaden the existing Arm product line' for use in computers, interactive TVs, PDAs, video games and digital imaging.[25]

Jaggar decamped to Austin, Texas, where DEC had a design centre, to work alongside the StrongARM team. On arrival, he wrote the first Arm Architecture Reference Manual, known as the Arm Arm, which was published in February 1996. 'They had an Arm just about implemented but we had never written down exactly what the architecture was . . . there were so many details that were just undocumented,' Jaggar said.[26] Realising he missed the warmer climate and that Arm needed to fish in a deeper pool of engineers, he made the move permanent, setting up the company's own processor design centre in the city.

But DEC had little time left to test the market with Strong-Arm. A lawsuit it launched in May 1997 determined the company's fate. DEC alleged that Intel's Pentium chip infringed several of its patents linked to its Alpha technology. The deep-pocketed Intel countersued.

Compared to its vast empire of old, there wasn't much left of DEC by this point. Since 1996, it had rid itself of its disk-drive operations, printer business and networking unit. To resolve matters, it might have been simpler for Intel to put DEC out of its misery. But because the FTC was investigating Intel's market dominance at that time it was keen to keep Alpha alive under separate ownership.

Hence an elaborate settlement was announced in October 1997: DEC kept Alpha, Intel paid DEC $700m and took over its Hudson, Massachusetts, manufacturing plant to produce Alpha chips, plus Samsung, IBM and AMD were granted Alpha licences. Regardless, the encouraging technology withered.

So did DEC. The complicated resolution opened the door for Compaq, by now the number-one PC maker, to acquire DEC for $9.6bn just three months later in January 1998, after two years of on-off talks.

Amid a flurry of corporate activity, the transfer of StrongARM to Intel attracted little attention, including, some observers speculate, within Intel. The company had worked on RISC-based processors before. Its i960 product line had enjoyed modest success but Intel stopped marketing the chip in its settlement with DEC.

Now it had a promising replacement that could be tailored for mobile phones and other handheld devices – even if the team that devised it did not hang around for long. For Robin Saxby, that meant further conversations not with Les Vadász, but Craig Barrett, Intel's chief executive.

Intel had a ready-made answer for device makers who would in future demand chips that put power consumption on a par with performance. But even though it would pour millions of dollars into StrongARM, relying on a small British company for a major market solution did not sit well with the American corporate titan. Intel's default position was to compete, not collaborate, as Arm would soon discover.

## Chapter 9

# SPACE INVADERS AND THE RACE FOR THE IPHONE

### The Grudge Match

When the phone call came in, Warren East's stomach lurched. Apple, the most celebrated user of Arm's chip designs, was requesting a conference call late that evening. As was often the case with the ultra-secretive Apple, East didn't know much more. But given what was going on in the marketplace, Arm's chief executive feared the worst.

It was 8 April 2008 and the British chip designer was riding high. The smartphone era had caught fire when Apple launched its iPhone a year earlier, and, just as they had powered Nokia devices for the previous decade, Arm designs were at its heart.

It wasn't just mobile phones Arm had to thank when its share price hit a 12-year high that spring. Arm had grown to become the largest microprocessor intellectual property company in the world, whose designs featured in Sony video cameras, Garmin navigation systems, Bosch car braking systems, Samsung disk drives and Toshiba televisions. Licensing to nearly every semiconductor company for use in everything from tiny, embedded microcontrollers to high-performance multicore processors had multiplied its scale. In January that year Arm announced the total number of processors shipped by its partners had passed 10 billion. Its success was celebrated at home, revered abroad – and made it a target for rivals to go after.

East and his executive team were spending a few days strategising about which industry segments Arm should focus efforts on in the long term. For the company whose engineers used to huddle in Swaffham Bulbeck's Black Horse pub after work, this offsite fitted its elevated status: Sopwell House, a four-star Georgian country hotel nestled in the greenery nearby St Albans in Hertfordshire and 50 miles south of Cambridge.

As mobile phones became smartphones, making calls and sending texts was giving way to emailing and web surfing – and watching movies on the move was coming soon. The mobile device was converging with the computer to become central to consumers' lives. That suited Arm, which saw mounting opportunity. Three-quarters of internet-connected devices used an Arm processor in their main chip, up from a quarter five years earlier. But this trend presented a grave threat to the chipmaker that still controlled the PC market: Intel.

One week earlier, Intel had made its move on mobile. At a developer forum in Shanghai, it announced that its Atom processor had begun delivery. Having warmed up the market for its arrival for months, the Atom, it said, represented a 90 per cent reduction in the power consumption of the x86 chip family as used in desktop PCs. Intel had built a decent relationship with Apple, supplying it with flash memory components, but the main attraction was partnering in iPhones on processor cores, just as it had begun to do for the Mac computer.

East knew that Apple was a powerful ally, capable of setting industry and consumer trends, but it could also be ruthless. Two weeks earlier, Wolfson Microelectronics, an Edinburgh-based chip designer, watched its shares slump as it disclosed that its audio chips that converted digital files into sound had been dropped from the next range of iPod music players. Despite its impressive momentum, Arm feared it could be next on the scrapheap.

Most of East's team were sent off to watch the football while, over a light meal, East and Mike Muller, Apple's key contact at Arm, prepared for the conference call. 'It was a horrible dinner, like a blooming funeral,' East said. 'We convinced ourselves that Intel had basically bought the business.'

On TV that night, Liverpool did battle with Arsenal in a dramatic Champions League fixture. Having faced off on the pitch three times in six days, here were two old foes who knew each other backwards. Both were intent on securing the ultimate prize, just like Arm and Intel.

## A Change of Leadership

Back in October 2001, Arm had swapped Robin Saxby's ebullient leadership for the more measured approach of Warren East. Saxby, who was knighted in the 2002 New Year honours, remained as group chairman for another five years.

It was a sign that the scrappy start-up of a decade ago was maturing, a sign that Saxby had done what he said he would do in establishing Arm's technology as an industry standard. It was time to pass the baton to a new chief executive.

Born in Aberdeen but raised in Usk, a small town in Monmouthshire, East was infused with a love of science by his Welsh mother, a chemistry teacher, and his Australian-born father, who worked as a lab technician for the local authority. Early on he divined that industry, not academia, was his career calling. 'I wanted to do business, not be in a white coat somewhere,' East said.

He completed an engineering science degree at Oxford University in 1983 but most of his practical experience came from a holiday job at Hoover's washing machine factory in Merthyr Tydfil.

There, East chopped up printed circuit boards with an old guillotine and etched copper off the boards in chemical vats. Persuaded that microchips were the future of electronics, he applied for work to the biggest maker of them in the world at that time, Texas Instruments.

Sixty miles north of London, TI's site in Bedford was its first outside the US when it opened in 1957, a year before Jack Kilby made his famous breakthrough. For many years, it was a fully integrated facility, not just fabricating microchips – with gold-wire bonding at first – but testing, packaging, designing and even making the silicon wafers in the same building.

East spent much of his time working on software for telephone systems, including a key component for British Telecom's Vanguard push-button phones, which is still used in millions of handsets today. 'I worked on the voice circuit and then wrote the software on the microprocessor that gave it all of its features and dialling,' he said.[1]

After studying an MBA at Cranfield University that broadened his leadership skills, East considered leaving. Instead, he was handed a choice commercial role as TI's European marketing director for field programmable gate arrays (FPGAs), integrated circuits containing logic blocks that could be configured for use after manufacture.

By spring 1994, it was harder to hang on. Career progression meant a move to Nice or Texas because TI had announced it would close the Bedford site that August. The company had licensed Arm the previous year so East, sensing potential, wrote to Saxby and asked for a job. He joined in September that year.

When he arrived, Arm had swelled to 70 staff but its revenues were not inflating at the same rate as the size of its workforce. East was tasked with setting up a consulting business to drive more revenues from helping licensees who were leaning hard on the business to design Arm processors into dedicated applications for them. His new colleagues 'could see they would never be able to devote enough resource to furthering the Arm technology roadmap', East said. 'All the resource was tied up supporting the customers who had bought licences.'

He marked himself out as smart, hardworking and honest, and was elevated to vice president of business operations in 1998. East, always neatly turned out, with thinning black hair and a precise manner, was also modest and thoughtful, withdrawing at the weekend to play the organ at his local church outside Cambridge and often called on to accompany weddings and funerals.

In 2000, he became chief operating officer and was elevated to Arm's board. It meant that when Saxby decided to step back East was a strong candidate to succeed him, even though that would mean leapfrogging a clutch of Arm founders, including Tudor Brown and Mike Muller.

'I guess I was less the charismatic type and more the operator who was going to install some scalability,' he said. East knew he had to hunt for new growth channels but was also keen to preserve some of the sensibilities of a small company and the close corporate culture Arm had bred. Every year, colleagues received from him a homemade Christmas card featuring intricate paper cut-outs, delivered on Christmas Eve. He might have become the suit he always aspired to be, but East still loved being hands-on. When a swimming pool was added to the family home outside Cambridge, he was the one at the controls of the mechanical digger shifting earth.

The heavy lifting extended to Arm. East could see that, for all it had achieved, Arm had tougher times ahead. The Arm-based baseband chip had been a stunning success, supplying Nokia and most of the mobile industry. But the company wasn't sure what came next, certainly not anything that delivered in such volume.

As far back as 1998, the year it floated its shares so successfully, Arm had been working on a project called 10X03 with the express purpose of growing the company's revenues tenfold by 2003. The first of the plan's two prongs was obvious – how to create a new microprocessor designs even more powerful and efficient than what had gone before. Secondly, Arm had to identify and meet the needs of customers all over again.

So far, its non-PC ambitions had built a strong franchise in mobile phones. That meant in 2002 the company passed an impressive milestone that demanded pause for thought: one billion chips had been shipped by its customers since Arm started out. But it was no time to get nostalgic or distracted by so many zeroes: now the hunt was on for the next billion, and the next.

Arm's marketers were convinced that part of the solution would come from phones once again, predicting internally there would be up to a dozen Arm processor cores per device one day. East was sceptical but knew that if the forecast came true there would be more royalties per device.

The company also needed to look further afield. It targeted 12 application areas, including TVs, cars and network controllers. Some opportunities, like digital TVs, were still way off, but automotive was encouraging. TI had used Arm chips in some anti-locking brake controllers, giving it a way into the industry. For Seagate, the American disk-drive maker, Arm developed a microprocessor that acted as a 'backbone' to its latest device, which worked remarkably similarly to a baseband modem.

East was upbeat about the long term but the here and now looked harder. The microchip market underwent one of its periodic busts in 2002, when exuberance among manufacturers meant supply rapidly outpaced demand, which could happen when consumers paused spending or there was a lull in product cycles. Arm was ejected from the FTSE 100 index in June as sentiment soured and nervous manufacturers held off from signing orders.

Some things didn't change, however. Sir Robin Saxby, once described by a Morgan Stanley analyst as 'the grit in the oyster that formed the pearl',[2] was as upbeat as ever about company prospects when he played tennis with some investment bankers at the Cliveden country house in Buckinghamshire that October. Little did he know then that three delayed deals would spark a profit warning a few days later that sent Arm shares tumbling.

## The Jesus Phone

The reboot of Apple began with the iMac, a futuristic, egg-shaped Macintosh computer with a 'Bondi blue' casing that was launched in May 1998. It was the first product to carry the imprint of Jony Ive, the British designer whose fondness for sleek aesthetics would blaze a trail through the consumer electronics world.

At the turn of the millennium, Steve Jobs was thinking bigger. The success of Nokia and others in popularising mobile phones threatened the personal computer as the centrepiece of consumers' digital lives. Efforts to reposition the PC as a 'hub', for collating music and images, explained the evolution of the iTunes service. Thoughts turned to creating Apple's own device for people to carry their music on – something better than the crop of digital players already on the market that carried just a handful of songs.

To lead development, the company hired Tony Fadell, a headstrong computer programmer and rock 'n' roll fan who had grown up listening to Led Zeppelin, the Rolling Stones and Aerosmith. He had pitched his ideas for music players to various firms, including his former employer, the Dutch electronics giant Philips.

The last portable device Apple had dreamed up was the Newton MessagePad and Apple still owned shares in Arm, whose designs lay at the heart of it. In fact, the proceeds from the staggered sale of Arm stock gave Apple some vital breathing space as it brought on new products – and new revenue streams. But in the rush to get what would become the iPod to market in time for Christmas 2001, once again employing Arm's technology appeared to have been purely coincidental.

Fadell hunted for a company whose work could provide a base for the new device and settled on PortalPlayer, a Santa Clara-based start-up that began life in 1999 when the venture capitalist Gordon Campbell funded a group exiting chipmaker National Semiconductor to create a chip system that could be used to power MP3 players.

Campbell was a former Intel marketing executive who had made a name for himself with his previous company, Chips & Technologies. As one of the first 'fabless' firms, its chipsets found a ready market among PC rivals eager to clone IBM's market-leading computer. The start-up was quickly bought by Campbell's alma mater, Intel, which supplied the IBM PC that others were eager to ape.

PortalPlayer was Campbell's gamble that another segment of the consumer electronics industry was about to explode. His investment firm, TechFarm, backed another company, N-able Technologies, which had created an access-control product using the ARM7TDMI design that had been utilised many times over since it had been incorporated into Nokia's breakthrough 6110 handset. To get off to a fast start, PortalPlayer adapted the design.

Whereas N-able used a single processing core, the PortalPlayer team judged it would need two cores with high-speed memory to ensure the songs it carried could be played back instantly. It worked with OKI Semiconductors, N-able's partner and an Arm licensee since 1995, to fabricate its first chip, the PP5001, but didn't deal with Arm directly.

A second version was designed and built by another firm, LSI Logic, which had enjoyed success supplying the microprocessor for the 32-bit Sony PlayStation. This late entrant to the console wars, which had been dominated by Nintendo and Sega up to that point, employed designs from RISC rival MIPS, but Portal-Player stuck with Arm.

When Apple came calling, PortalPlayer was ready for pro-duction and going through the design development process with Sony and Panasonic. Fadell moved faster, and soon asked for exclusivity. It meant that Campbell struck gold for the second time and Arm found its way into another Apple device, this time one that would fare better than the Newton.

For the project dubbed 'Crossover', Apple's in-house designers were focused on perfecting the look and feel of the iPod, includ-ing the wheel that consumers would use to scroll through their song playlists. Without its own specialist chip designers, it relied on PortalPlayer to contribute the innards, including the iPod's operating system for audio playback that capitalised on Arm's low-power architecture.

Even though it could carry 1,000 songs, music fans quibbled at the $400 price tag of the iPod and the fact that initially it was not compatible with Microsoft Windows PCs. To transfer music to the device, they needed to use a Mac. But the iPod's eventual success, creating a new category of device as a direct descendent of Sony's Walkman, changed the perception of Apple. It filled its developers with confidence and allowed them to dream about what came next.

---

A phone was obvious. Just like music players, Jobs saw a market that could be measurably improved upon. And, as phones began to incorporate their own music players, it was a defensive move for a company that was selling millions of iPods. Having learned from the failed collaboration with Motorola on a music phone, ROKR, Apple's own phone project gained internal approval at the end of 2004 and was given the codename Project Purple.

Terry Lambert, a senior software engineer, was brought into the project late on for 'debugging', that is, ironing out glitches found in the phone's operating system. Even though he was a member of staff, there were a series of non-disclosure agreements to sign. 'You couldn't see the code name, until you agreed not to discuss the code name,' Lambert wrote in a 2017 blog post. Inside the development area, black cloths covered everything. 'I only got to see the machine doing the remote debugging, not the target — but it was obviously an Arm-based system,' he added.[3]

The work took place in absolute secrecy in the Mariani 1 building on Apple's campus in Cupertino, California, where security patrolled the main entrance and individual pass keys were mandatory. At their computer screens, engineers used electronic design automation (EDA) tools to capture the schematics and printed circuit board layout from which prototypes could be built.

Development was riven with disagreement. 'Steve loved creative tension and stoking that fire a little bit,' remembered David Tupman, a hardware engineer who joined Apple just as the iPod was readying for launch after cutting his teeth on Psion's electronic organisers in the UK. About the only thing the two factions pushing different approaches for the device didn't disagree on was the choice of processor architecture.

On one side there was Scott Forstall, who had worked under Jobs at NeXT, his computer company created during the wilderness years away from Apple. Forstall led a team determined to squeeze the Mac's operating system into a small tablet with touch controls that could make calls. His colleagues thought Arm's technology could support what would effectively be a stripped-down computer. 'We knew for sure that there was enough horsepower to run a modern operating system,' said Richard Williamson, a senior software engineer and fellow NeXT alumnus.[4]

The other side was led by Fadell, who sought to splice his creation with a phone. Adding phone capability to the iPod was something East's team had promoted to Apple too, after its designs had been adopted across the MP3-player market. Apple needed to be sure that Arm could keep delivering, but the fact that Doug Dunn, Fadell's former boss at Philips, was an Arm board member and would become chairman in October 2006 did relations no harm at all.

The iPod team favoured using a version of Linux, the open-source operating system, which just happened to run on Arm as well. However, as the debate went on another supplier fancied its chances at getting into the iPhone: Intel.

———

Less than three weeks after becoming chief executive, Paul Otellini celebrated a significant coup. On 6 June 2005, Apple agreed to switch its Mac computers to Intel chips for the first time. To please the most demanding customer in town, it seemed that Otellini's efforts to be more user-friendly were working.

Since its debut in 1984, the Mac had run on Motorola and then PowerPC chips, but the alliance Apple had forged with IBM and Motorola to design its own silicon was no longer delivering. Jobs' choice to go with Intel marked an extraordinary truce after years of animosity. When Apple unveiled its Power Macintosh G3 in spring 1998 it was backed by an advertising campaign that pictured a Pentium II chip on a snail's back in one commercial and, in another, a singed Intel dancer being dampened down by a firefighter. In the intervening years, concerns over the speed and power consumption of Intel's chips had obviously been extinguished.

Otellini marked the partnership by joining Jobs on stage at the Macworld event on 10 January 2006 at San Francisco's Moscone Center, emerging from a cloud of dry ice, wearing a bunny suit and delivering a shiny silicon wafer disc. 'It's been incredible how our engineers have bonded and how well this has gone,' Jobs said after they greeted each other.

The Intel boss described the process, which had involved over a thousand of his staff, as 'energising, challenging, fun'. It reminded him of a quote from the company's co-founder, Robert Noyce: 'Don't be encumbered by history, go out and create something wonderful,' he told Jobs. 'Well, we did, and we did it together.'[5]

The switch was hugely complicated but ultimately seamless. The bonding emboldened Intel to think it could partner Apple on its phone project too and power its way into the mobile mass market. But Apple's engineers were not so sure. Power consumption was a concern, but so was price. 'Back then Intel couldn't ship an empty box for $20,' said Tupman. 'We needed a chip for less than $20.'

———————

In the end, Apple turned to Samsung and the pair struck up a relationship in 2005. For subsequent iPod editions, Apple had opted to ditch the hard disk drive. It needed a reliable supplier of lightweight and efficient NAND flash memory chips to provide permanent storage – and it needed them in great volume. PortalPlayer had exceeded its usefulness. Now the South Korean electronics giant fitted the bill.

Flash memory was an area that Samsung had identified for growth soon after the collapse of the market for DRAM chips that were used for temporary storage. Making a strong statement of intent, it based its first fab outside South Korea in Austin, Texas, in 1997, with an investment from Intel in exchange for a guaranteed supply of memory chips.

Hwang Chang-Gyu, who went on to lead Samsung Electronics, recalled travelling to meet Steve Jobs in Palo Alto to discuss his needs. 'I met him with the solution to Apple's life-or-death problem hidden deep in my pocket,' he said.[6]

Inspired by Moore's Law, he had coined the term 'Hwang's Law' to support his prediction that memory density would double every year. Samsung engineers set about proving Hwang right and the outcome paid dividends.

The relationship between the two firms could trace its roots earlier, to when Jobs visited B.C. Lee in Suwon, just south of South Korea's capital, Seoul, in late 1983. It was just before the Macintosh was unleashed on the world and Jobs was already looking ahead, throwing around the idea for a tablet computer in the mould of the Dynabook concept – and working out what components he would need to make it a reality.

For Lee, it was nine months after his Tokyo Declaration and he was just beginning to supply memory chips to manufacturers. Despite the 45-year age gap and Jobs' apparent failure to show due deference to his elders, they got on well, with Lee later declaring to his assistants, 'Jobs is the figure who can stand against IBM,' the dominant PC maker of the time.[7] That didn't look immediately possible when Jobs departed Apple under a cloud two years later.

Samsung was no stranger to mobile phones. Its first handset was launched in 1988, the year after Lee's death, although it didn't sell them outside South Korea until 1996. The product line was a natural step forward for a conglomerate that had made consumer electronics for decades and its own microchips since 1974. In 2007, the year that the iPhone launched, more than one billion phones were sold and Samsung was third in the market behind Nokia and Motorola with a 13 per cent share.[8]

The fact Samsung made both microchips and handsets should have been a great advantage for the business. But the two separate divisions had a notoriously fractious relationship. The consumer-electronics operation viewed its semiconductor sibling as just another supplier and went elsewhere if the price and quality were not deemed good enough. That suited Apple, which gained assurances about Samsung's internal firewall.

When it came to launching the iPhone, Apple had been toying with using an Arm-based video chip from Broadcom but was hunting for something better. When Samsung executives came in for their quarterly review in summer 2006, they were quizzed on whether they could produce an Arm chip with a graphics controller and fast memory in a hurry.

Hwang despatched a team to Cupertino to build a new chip alongside Apple's engineers. 'Normally it takes days per layer, and it's twenty or thirty layers of silicon you're trying to build,' said Tupman. 'Normally it's months and months to get your prototypes. And they were turning this around in six weeks. Crazy.'[9]

———

And so it was that powering new innovations such as a touch-screen that could be manipulated with finger pinches and zooms and damage-resistant 'gorilla' glass was a chip brain adapted from a cable box used for a DVD player.

On 9 January 2007, Jobs walked on stage at the Macworld San Francisco event. 'This is a day I've been looking forward to for two and a half years,' he said, to whoops and a smattering of applause. 'Every once in a while, a revolutionary product comes along that changes everything,' Jobs, dressed in jeans and trademark black rollneck with his sleeves rolled up, told the audience.

Reminding them of Apple's past hits – the Macintosh in 1984 and 2001's iPod – he went on: 'Well today we're introducing three revolutionary products of this class. The first one is a wide-screen iPod with touch controls. The second is a revolutionary mobile phone. And the third is a breakthrough internet communications device.'

The crowd were excited but didn't quite comprehend what he was saying. Jobs repeated himself and then began chanting. 'An iPod, a phone, and an internet communicator . . . Are you getting it? These are not three separate devices, this is one device,' he said, drifting to centre stage. 'And we are calling it iPhone.'[10]

As the crowd erupted, Arm was nowhere to be seen. Nevertheless, its fortunes were about to be transformed by Apple once again as what was dubbed the 'Jesus phone' quickly inspired religious fervour.

## Tanks on Intel's Lawn

It wasn't just Intel trying to drive into what was seen as Arm's territory. Arm was already heading in the opposite direction.

More than two years before the iPhone launch, when Warren East pulled back the curtain on the first day of the inaugural Arm Developers' Conference (DevCon), all he could see was Intel.

Out of the window on the morning of 19 October 2004, East gazed at Intel's 84,000 square metre(900,000 square foot) fabrication plant, known as D2, the last large-scale fab in Silicon Valley, where new products such as flash memory chips for cellphones and digital cameras were put into pilot production. Behind it, on Mission College Boulevard, lay the microchip giant's towering global headquarters with its ice-cool blue glass frontage.

Arm's first partner meeting more than a decade earlier was a homely affair. It comprised a week of discussions for licensees that spilled out into picnics on the lawn by the barn in Swaffham Bulbeck and continued in the local pub every night.

In contrast, this first DevCon resembled a village in its own right, welcoming over three days more than 2,000 people, twice initial estimates, with almost a hundred companies exhibiting. And it was taking place in Santa Clara, Intel's backyard, at the very conference centre where Andy Grove would bid farewell to adoring shareholders the following May. 'I do remember feeling a real tinge of excitement that here we are, we've parked the tanks on their lawn,' East said.

Harking back to Arm's early penny-pinching, he had only sanctioned the event to excite developers about their future vision and build community support on the proviso it cost the company nothing. Rather than rolling into town with the purpose of baiting its biggest rival, the location had been decided by a conference organiser eager to turn a profit.

On stage that day, East unveiled the M3 processor, the first of Arm's Cortex 'family' that would launch over the next three years. To some extent, Cortex was simply branding applied to Arm's existing product roadmap, a way to bring the 10X03 strategy to life. The company had fallen well short of its ambitious tenfold revenue increase by 2003, but the plan had helped to define the business areas it needed to pursue. Arm said Cortex was 'more clearly segmenting its products in order to satisfy the demands of more and more diverse markets'.[11] More brands and more segments would follow over the years.

The announcement also put down a marker. The M3 was 'specifically designed to meet the requirement for high system performance in extremely cost-sensitive embedded applications, such as microcontrollers, automotive body systems, white goods, and networking devices', the company said. Thanks to Nokia and TI, Arm technology had become an essential contributor to almost every mobile phone on the planet, but now it was broadening its horizon. In doing so, it could not help increasing its focus on Intel, keeper of the world's only other microprocessor family of scale, the x86.

---

Two months before DevCon, Arm announced its first significant acquisition. News on 23 August 2004 that it was paying $913m for Silicon Valley's Artisan Components was poorly received by investors, who sent the shares down 18 per cent on the day.

Alongside Arm's microchip work, Artisan owned a library of basic designs that helped to integrate software with electronic circuits, which was increasingly important as chips became more complex and manufacturing harder. 'If you were to compare the chip industry to car manufacturing, Arm designs the engine and Artisan designs the pistons,' said Sir Robin Saxby, Arm's chairman, as he tried to explain the rationale to doubting media and shareholders.[12]

In retrospect, some Arm executives regarded the Artisan deal as the right idea at the wrong price – and perhaps it was even the wrong target too. It certainly offered Arm a new channel for growth. By adding Artisan's products to its own, East said he hoped over the next two to five years to double the royalty of about nine cents the company currently received for every Arm-powered chip sold.[13]

What few realised – and the company did not acknowledge publicly – was that Arm viewed the deal as an insurance policy in its long-term face-off with Intel. Looking out five years, Arm executives saw that Intel's integrated model of chip design and manufacture would give it greater competitive advantage as silicon geometry shrank. Executives thought that adding a library company would help its customers, many of whom relied on foundries for manufacturing, to tighten the link between microchip creation and construction. Selling abstract microprocessor design was no longer enough: it had to think about implementation too.

———

Despite the strategic manoeuvring, Arm continued to work with Intel on StrongARM, the technology that the American firm had inherited in the settlement with Digital Equipment Corporation (DEC) in 1997. The relationship was cordial, not dissimilar to how Arm worked with any other semiconductor partner. From time to time, East would meet Sean Maloney, Intel's executive vice president and the general manager of its mobility group. The shaven-headed Londoner was regarded as chief executive Paul Otellini's de facto deputy.

With the introduction of a new microprocessor core, Strong-Arm was rebranded as XScale in 2000. It had little success getting into cellphones but did better with handheld computers such as the Palm Treo, Research in Motion's BlackBerry and HTC devices. Still, Arm directors had their reservations about Intel's commitment to the partnership.

In February 2005, at the mobile industry's big annual gathering, the 3GSM World Congress, the industry was in bullish mood as it geared up for a second wave of growth. At the event in Cannes, Nokia predicted that global sales of 3G mobile phones would soar from 16 million to 70 million that year, as part of a jump in overall phone sales from 1.7 billion to 3 billion by 2010. Millions of people in developing markets who had never known landline telephony were being connected for the first time by cheap, simple handsets made possible in part by low-cost, low-power microprocessors.

But when East and Maloney met behind the scenes, the tone was far less celebratory. 'We are not going to do anything more with XScale,' East recalled Maloney saying. The sudden change of stance set East's mind racing. If Intel wasn't going to work *with* Arm to grow its mobile business, it was surely going to work against it. Battle lines had been drawn.

---

On 27 June 2006, despite billions of dollars invested, Intel announced the sale of its communications and application processor business, including XScale, to Marvell Technology Group for $600m. It was left to Maloney to justify the eye-catching reverse. 'The communications and application processor segments continue to present an attractive market opportunity, and we believe this business and its assets are an optimal fit for Marvell,' he said in a statement. 'We have a long history of working closely with Marvell and believe it has the ability to grow the business while maintaining customer commitments.'[14]

In a later interview, Maloney talked of a 'budget crunch' when faced with investing in another version of XScale or Banias, Intel's latest x86-based Centrino microprocessor. 'We didn't have the headcount to do both,' he said. 'We realised we had a greater commitment to x86 so we shifted resources to that.'[15]

Even though Arm, with its low-cost, low-power designs, had energised the mobile revolution so far, and Intel had a high-performance version of that technology in the palm of its hand, it was dropping it. Instead of smart, handheld devices, it was focusing back on its core PC business, with an eye on netbooks, as well as other emerging technologies for mobile computing. As the market coalesced, that sounded like nothing more than semantics.

For Paul Otellini, it was a stark choice. He knew that even a vast company like Intel would struggle to sustain two computer architectures. It confused customers and had the potential to set its own developers against one other. The company had briefly tried to bring on a second 'language' before, toying with the i860, a RISC microprocessor design, as long ago as 1989.

'Should we abandon a good thing, which for now at least was a sure thing, and lower ourselves back down into a competitive battle with the other RISC architectures, a battle in which we had no particular advantage?' Andy Grove debated years earlier in his book *Only the Paranoid Survive*, before recommitting to the CISC architecture.[16]

In Otellini's time, the x86 was still supremely profitable, a priceless, entrenched franchise in which billions of lines of code had been written. If it could possibly avoid it, Intel had no intention of compromising that by hedging its bets and powering Arm's value higher.

The competitive pressures showed up in industry data. Even Intel's previously rock-solid market share in PCs was being nibbled away, down from 83 per cent to 75 per cent in the six years to 2006 according to market tracker IDC, largely because of resurgent rival AMD. The only option was to double down on x86.

Even though Arm had captured the iPhone opportunity, Intel had the Arm board worried. Margins were strong and customers were plentiful for the British firm, so little time needed to be given over to operational issues at board meetings. That meant time set aside to talk strategy – and that meant Intel-watching while newspapers and websites speculated over the chip giant's next move.

Directors were sufficiently experienced to have seen what Intel did to build and hold a near-monopoly in the PC market, including the battles with AMD. With its vast financial power, they feared Intel could yet combine its architectural knowledge and market credibility to destroy the solid base Arm had built.

'For maybe two years that was the primary concern of ours,' said one Arm board director from that era. 'There was no board meeting in which we didn't have a quick review. Where are Intel? What have they got? What are they doing with it? Are any customers showing any serious interest? It could have been make or break for Arm if Intel had really powered ahead.'

## The Licence

In late 2007, as iPhone sales ignited around the world, speculation mounted that Intel had made a breakthrough. Apple Insider, a well-informed website covering all things Apple, reported that the company would form a 'closer bond' with Intel early in the New Year 'when it begins building a new breed of ultra-mobile processors from the chipmaker into a fresh generation of hand-held devices'.[17]

The news suggested a great result for the 'Apple group' of engineers and sales staff that Intel had set up to win the company's business. Months earlier, the group's director, Deborah Conrad, had told reporters that Apple's way of looking at the world was making Intel 'think different' about its own business and, when it came to the prospect of future gadgets, her team got 'very, very excited'.

The website suggested two contenders for the new partnership, either a next-generation iPhone or an 'ultra-portable slate computer'. And the chip to be used would be Silverthorne, Intel's new design for cellphones and other mobile devices that Otellini had ranked in importance alongside its legendary 8088 processor or Pentium in an interview in June that year.

A few weeks after the report, Apple debuted the MacBook Air, a slimline notebook computer, dubbed an 'ultrabook'. It was equipped with an Intel chip, contained in a new, miniaturised package, so it took up less space, but it wasn't the chip or the device the report was hinting at.

As related in his biography written by Walter Isaacson, Jobs had planned to use Intel's low-power Atom chip in the iPad, his 'slate computer', but his engineering team had other ideas. Fadell, a key lieutenant, favoured something based on Arm's designs, which were much simpler, more energy-efficient and had featured in the original iPhone.

'Wrong, wrong, wrong!' Fadell shouted in one highly charged meeting where Jobs argued it was better to trust Intel.[18] Fadell even placed his Apple badge on the table, a sign he was prepared to quit over the issue.

The outside world discovered what all the fuss was about at a developer forum in Shanghai on 1 April 2008, when Intel announced it had begun shipping its Atom processor, which had been rebranded from Silverthorne. The Atom represented a 90 per cent reduction in the power consumption of the x86 chip family as used in desktop PCs – although it still used more power than the Arm equivalent. It also featured a 'drowsy transistor' that would power down to save energy when the chip was not computing.

Intel's pitch was that it could successfully translate PC performance into a mobility internet device (MID), as it insisted on calling this category of simple web-surfing machines. What it was clearly not was a chip designer for basic handsets straining to support phones that were fast transforming into computers.

Investors were dubious. They were not sure if Intel was properly addressing the PC's collision course with the high-end cellphone. And they didn't know how the Atom could avoid cannibalising Intel's existing market.

For Apple, the attraction of using Atom was compatibility with the line of x86 processors that it had begun using for its Macs. But that wasn't enough to swing the deal.

———————

Of course, Warren East wasn't to know Apple's verdict when he joined the evening conference call from Arm's Sopwell House retreat just one week after Intel had proudly debuted its Atom chip in 2008.

With East fearing the worst, instead Apple relayed the best news possible. The tech giant didn't just want to continue using Arm's designs in its growing array of electronic devices, it wanted to build a deeper relationship with Arm as it did so.

The company requested an architecture licence from Arm, which meant it could significantly vary its own designs from Arm's blueprints rather than just license the standard building blocks. It was a key plank in Apple's plan to take greater control of its own silicon, of which more later.

Arm had only granted a handful of such licences in the past, and all to chip companies. Given the extra flexibility, East could have charged much more. But Arm was too delighted to play hard-to-get on price.

Ultimately, Apple concluded that Atom was not efficient enough to power the iPad. And Intel's approach caused consternation as well, judged too rigid in how it worked, and eager for Apple to conform with the way it built and tested chips. Apple opted for Arm, with Samsung once again manufacturing at its fabs in Giheung, not far from its Suwon headquarters south of Seoul. Notably, the A4 chip it debuted with the iPad on 27 January 2010 was the first Apple had designed itself.

'We tried to help Intel, but they don't listen much,' Jobs said. 'We've been telling them for years that their graphics suck. Every quarter we schedule a meeting with me and our top three guys and Paul Otellini.' The relationship started well. 'At the beginning, we were doing wonderful things together. They wanted this big joint project to do chips for future iPhones.'

Jobs went on: 'There were two reasons we didn't go with them. One was that they are just really slow. They're like a steamship, not very flexible. We're used to going pretty fast. Second is that we just didn't want to teach them everything, which they could go and sell to our competitors.'[19]

———

Arm expected to feel the iPhone benefit eventually. At the company's annual results presentation on 5 February 2008, East reported to investors that it was supplying an average of 1.6 microprocessors per Arm-based phone, up from 1.2 in 2003. The forecast that that number would trend up to two per phone was taking longer to fulfil, East said, because 'there has been superior growth in the overall number of phones out there, driven quite substantially in 2005 and 2006 by explosive growth in low-end handsets'.[20]

So far in the smartphone category, Arm had 100 per cent market share, and because it was supplying special-purpose processor designs, such as those that handled graphics, as well as general-purpose chips, 'we don't just earn $0.05 or $0.06' in royalties per unit, 'we earn $0.50 or $0.60', East added.

By the end of 2010, and less than three years after the iPhone launched, the smartphone market was already bigger than the PC market based on units sold, according to IDC.

However, Arm still thought it was worth travelling in that direction. Gaining traction in the laptop market was not easy but the company sensed an opening when Taiwan's design manufacturers – which built products to be sold under their customers' brand – ventured into creating their own designs.

Arm also decided to target another of Intel's near-monopolies and its most profitable market, computer servers, which it had grown to dominate despite an early scare from AMD and its Opteron processor. If it could grab a 10 per cent share, directors reasoned, that would be enough to exert some pricing pressure over Intel and perhaps throttle back its research spending. To ensure that his colleagues were all focused on a common enemy, East ensured pictures of the Intel leadership team were pinned up on the wall at meetings of his senior management.

Much depended on how Arm's partners felt about its capabilities. Intel was not just competing with one firm, but Arm's 200 licensees, from Qualcomm to TI. In an interview with *Fortune* magazine, East paid tribute to Intel as a brilliant competitor and a leading manufacturer. But he added: 'Intel's got an old-fashioned business model. Arm's got a 21st-century business model.'[21]

---

As soon as the first iPhone was released, Apple had stepped up its relationship with Arm, and the architecture licence only firmed it up further. The tech giant was keen to get to know its supplier's supplier and the wider ecosystem it relied upon. With future models in mind, it needed chips that ran on reduced power, but that also enabled improved performance.

A senior team of up to 15 people, both managerial and technical, fell into a rhythm of spending one week in the UK twice a year. The likes of Tony Fadell and Apple's senior vice president Bob Mansfield would base themselves in London, often at the Courthouse Hotel, a five-star venue converted from a Grade II listed magistrates' court just opposite the Liberty department store off Regent Street.

Days would be spent travelling out of London to see Arm in Cambridge, as well as the supplier of the iPhone's graphics processor, Imagination Technologies, which was based in Kings Langley, just north of Watford. At night, Tupman made sure his colleagues saw London's sights, often taking in a West End show or, on one occasion, travelling down to the dog track in Wimbledon for dinner and a surprisingly competitive race night. When he left Apple in 2011, he was vice president of iPhone and iPod hardware engineering, running a 200-strong team that worked on the devices' electronics, audio, camera and more.

The Arm meetings were good-natured. East never met Jobs, who is not thought to have visited Cambridge either, but Apple viewed the relationship as a partnership. That meant understanding the full picture: not just the processor core, but all the peripherals the Arm team was working on.

The only area where the Brits disappointed was with their catering. The American visitors could not abide the trays of forlorn sandwiches they were served for lunch most days.

## Retired Hurt

When Paul Otellini retired from Intel in 2013, the numbers spoke of a giant leap forward. Compared to the moment when Grove had taken his leave eight years earlier, in 2012 the company generated revenues of $53.3bn – an uplift of 56 per cent – and net income of $11bn, some 47 per cent higher. Shareholders had felt the benefit in their pocket too, as the annual dividend swelled fourfold to $4.4bn.

The fight to maintain market leadership had, predictably, become more complex and more expensive but the technology was still marching relentlessly onward. Over the period, Intel had stepped up delivery of its 22nm processors – employing geometry a third of the size of that was being used the start of Otellini's tenure – and, looking ahead, factories capable of operating 14nm were being built and chips 10nm and smaller was in development. Estimating that its process technology still commanded a two-year lead over the competition, it wasn't hard to imagine where Intel's annual $10.1bn research and development budget – more than twice the 2004 figure – was going.[22]

But for all the bullishness and big spending, Otellini's time was forever marked by what he didn't do, and that appears to have coloured the manner of his exit. Unlike his four predecessors, he did not step up to chair Intel. In fact, he departed a couple of years before the company's retirement date of 65 dictated he should.

Apple's iPhone had flown. In 2012, its fifth birthday year, some 125 million units were sold, generating an astonishing $80bn in revenues, which equated to more than the entire Microsoft business could muster. There were not enough superlatives.[23]

Consumers' lives now revolved around handsets that contained their calendar, photos, social media, games and music. Nokia and Research in Motion, maker of the businessman's one-time favourite BlackBerry, were in sharp decline. And PC makers including HP and Dell were struggling to respond. Apple could stake a claim at having formulated the most successful and disruptive product in history.

In a mea culpa interview on his last day in the office, Otellini spoke about his failure to install Intel chips in the iPhone. No one knew what the device would become when Apple insisted it was prepared to pay a certain price for the chip 'and not a nickel more', he said. 'And in hindsight, the forecasted cost was wrong and the volume was 100x what anyone thought.'[24]

With regret, he added: 'The lesson I took away from that was, while we like to speak with data around here, so many times in my career I've ended up making decisions with my gut, and I should have followed my gut . . . my gut told me to say yes.'

He was inevitably quizzed on Arm. 'Arm is an architecture. It's a licensing company,' Otellini said. 'If I wanted to compete with Arm, I'd say let's license Intel architecture out to anyone that wants it and have at it and we'll make our money on royalties. And we'd be about a third the size of the company.'

––––––––––

Otellini was right. In the same year that Intel revenues were $53.3bn, Arm reported a more modest $913m – one-sixtieth of its rival – and a pretax profit of £277m as the average royalty fee it commanded for every chip its customers sold nudged up to 4.8 cents in the fourth quarter.[25]

The 16 per cent rise in annual sales spoke of growth significantly faster than the wider semiconductor industry, good demand for its Cortex-A class processors and Mali graphics engines and gains in consumer electronics, including digital TVs. But far from doubling the royalty rate, as East had hoped at the time of the Artisan deal in 2004, it had almost halved thanks to an explosion in lower-value devices connected by Bluetooth and wi-fi that pointed the way to the coming 'internet of things' (IOT) trend.

And for all its hunt for new destinations for its designs, the mobile market still dominated Arm's income, where it earned a sliver of Apple's fortune. Its designs featured in an estimated 2.5 chips per iPhone and, because they were complex, Arm could charge double its average royalty rate.

What began with the baseband modem, the chip that communicated with the telephone network, expanded to include the apps processor – which talked to the user – the phone's applications and its display. Over time Arm designs featured in chips that ran wi-fi, Bluetooth, location-based services and the camera. These functions typically started as separate chips and then were integrated, but they were most easily integrated as a separate Arm processor and subsystem, therefore generating extra royalties.

For 2012, Apple still only added up to an estimated £35m of Arm's profit, probably less than Samsung contributed that year. Still, its modest cut was the price of ubiquity and the brand pull of Apple was terrific. It ensured Arm could make the transition from 'dumb' phone to smartphone that eluded so many well-known handset brands, including, in the end, Nokia, which sold its devices business to Microsoft for $7.2bn in September 2013.

What Arm lacked in financial might it made up for in influence. Some 8.7 billion Arm-based processors were shipped in 2012 – the equivalent of 70 chips for every iPhone sold. Precisely because its technology was inexpensive, licensees did not need to consider an alternative and the Arm family spread and spread. In the battle for mobile leadership, a loose alliance of developers had won out over Intel's imperial invincibility.

At least by concentrating on x86, Intel channelled its research spending, which worked well for customers. Their ability to reuse software because of some level of compatibility with chips going back five decades to the 8008 meant they could get product to market faster, with fewer bugs.

Otellini said his staff should be focused on competition from Qualcomm, TI or the graphics specialist Nvidia, but not Arm, 'or if someone like Apple is using Arm to build a phone chip, I want our guys focused on building the best chip for Apple, so they want to buy our stuff'.[26]

---

On 26 April 2016, Intel announced it was effectively ditching Atom, its great move into mobility, as well as cutting 12,000 jobs as the PC market declined. It appeared that, in mobile at least, it had been outflanked just as the PCs it powered did for DEC's minicomputers in the 1980s. Intel and DEC both doubted the profitability of what came next. They also both underestimated the volumes that ultimately made up for thinner margins.

Otellini's successor as chief executive, Brian Krzanich, said his strategy was about 'transforming Intel from a PC company to a company that powers the cloud and billions of smart, connected computing devices', adding that: 'We are in a time when technology is valued not just for the devices it produces, but for the experiences it makes possible.'[27]

The market for chips that powered computer servers was the next battleground on which Intel and Arm would meet. But, illustrating the incestuous nature of the microchip industry, Intel also announced it would begin manufacturing Arm-based smartphone chips for customers, starting with LG.

Four years after Krzanich's pronouncement, on 22 June 2020 Apple's chief executive Tim Cook announced 'a historic day for the Mac'.[28] The effect of Otellini's great charm offensive had expired. The technology giant was dropping Intel chips from the Mac in favour of 'Apple silicon', its own-house chips developed on Arm architecture and a direct result of that surprise 2008 conference call. 'Apple silicon will make the Mac stronger and more capable than ever,' added Cook. 'I've never been more excited about the future of the Mac.'

For Apple, the move gave it greater control over its supply chain and made it easier for apps developed on iPhone, iPad and the iOS operating system to work on the Mac without modification. Its strides in the area of 'custom silicon' had profound implications for the industry that had long separated chipmakers from chip consumers.

It was also a historic day for Arm, and another twist in its long-standing relationship with Apple. By confirming its use in laptops, beyond mobile phones and smartphones, it was extending from one market to another in a way that Intel had not so easily managed in the opposite direction. Drawn in part by the superior battery life it offered, Microsoft followed suit soon after. The software giant had first released a mobile operating system that ran on Arm in 2011.

## Chapter 10

# A 300-YEAR VISION, A 64-DAY TAKEOVER

## Lunch on the Waterfront

The Pineapple restaurant in Turkey's historic port city of Marmaris had enjoyed brisk trade since it first opened in 1988. From morning till night, it served sailors and tourists a mixture of fish, grilled meats, pizza and pasta. Diners with a large appetite could attempt one of the chef's signature dishes: a baked pineapple stuffed with seafood.

The restaurant was situated halfway along the waterfront overlooking the marina. Fringed with palm trees and unruly purple bougainvillea, it was well known as a lovely spot to unwind under the yellow awning and look out over the hundreds of boats bobbing in the bay.

As a gateway to the eastern Mediterranean, Marmaris had a walk-on part at significant moments in history. Lord Nelson and his fleet sheltered in the harbour here in 1798, en route to Egypt to defeat Napoleon's armada. Suleiman the Magnificent fortified Marmaris Castle, which stands on the hill overlooking the sea, as a base for the Ottoman navy during the five-month siege of Rhodes in 1522.

On 3 July 2016, Marmaris was the unlikely scene for a notable moment in corporate, not naval, history. That day, the Pineapple's first floor was booked out entirely by a single party.

There were four of them: Simon Segars, who in 2013 had become Arm's third chief executive; his chairman, Stuart Chambers; Masayoshi Son, the billionaire Japanese investor whose portfolio of investments included Chinese ecommerce website Alibaba and the US mobile-phone carrier Sprint; and Alok Sama, chief financial officer of Son's SoftBank investment company.

They ordered lunch and chatted, like four friends relaxing on a summer Sunday afternoon. But for this hastily arranged meeting there was really only one thing on the menu: the future ownership of Arm, the UK's most successful technology company.

## Animal Instincts

Masayoshi Son had built a reputation for taking quick decisions. When he first met Jack Ma, the founder of the Chinese ecommerce website Alibaba, it took a matter of minutes for him to decide to invest. It was October 1999 and the first internet bubble was substantially inflated, but such was the excitement about Chinese technology stocks few investors were exercising caution.

The meeting in Beijing was among several brokered for Son by Goldman Sachs, the investment bank that a few weeks earlier had led a $5m funding round in Alibaba, valuing the venture that had only been set up that March at $10m.

Son and Ma instantly saw something of themselves in each other. Both had humble backgrounds, both spoke their minds, both had outrageous ambition – and modest stature. 'We didn't talk about revenues; we didn't even talk about a business model,' Ma said in one account of the meeting. 'We just talked about a shared vision.'[1]

Soon after, Ma came to Tokyo to talk terms. Son immediately offered him $20m for a 40 per cent stake in the business, which would mark a threefold uplift for Goldman and partners in just a handful of weeks. When Ma demurred, not wanting to give away so much equity, Son doubled his offer. They later settled on $20m for a 30 per cent stake. Explaining his decision years later, Son said, 'it was the look in his eye, it was an animal smell', that encouraged him to splash the cash so readily.[2]

Announced on 18 January 2000, the $20m syndicate led by SoftBank, along with the promise of extra capital to develop international language versions of its websites, turned out to be vital to see Alibaba through the imminent internet bust. For Son, whose portfolio crashed in value, it became the investment that underpinned all else as he battled back. When Alibaba listed its shares in New York in September 2014 the company raised $25bn, making it the world's biggest initial public offering. SoftBank's 32 per cent stake was valued at a cool $75bn and Son was confirmed as Japan's richest man. From 100,000 members in early 2000, Alibaba had grown to house 8 million sellers of goods and had peak one-day shipments of 156 million packages during a single day in 2013.

---

That initial investment was typical of the impulsiveness that ran through Son's rags-to-riches tale. His grandparents had come to Japan from South Korea hiding in the hull of a fishing boat. The family had nothing and lived in a shanty town on land belonging to Japanese National Railways. Son's father was a pig farmer who brewed illegal alcohol on the side but hauled himself up to enjoy success with restaurants and parlours specialising in pachinko, a Japanese version of pinball.

In the summer of 1973, when he was 16, Son's family was wealthy enough to send him to California to be educated. So the story goes, on a shopping trip to his local Safeway grocery store between language classes, he was flicking through a copy of *Popular Electronics* magazine when he happened upon a close-up picture of Intel's newly announced 8080 chip, the antecedent of the x86 architecture that would come to dominate personal computers. Son related that it left a deep impression on him.

'It was like when you watch a really powerful scene in a film or get caught up in a piece of music and you start tingling all over,' he said. 'It was the exact same feeling. I was tingling all over and warm tears started falling down my face.'[3]

Son bought the magazine, cut out the picture and put it in a clear plastic file to store in the rucksack that he always carried with him. That included sleeping with it under his pillow at night. He decided he had to get involved in the computer industry.

Son was admitted to study economics and computer science at the University of California, Berkeley, in 1977, where students could take advantage of rows of computers, available 24 hours a day. The entrepreneurialism began while still on campus when a speech translator he designed was licensed to Sharp, the Japanese electronics firm. Then Son imported games consoles from Japan to the US.

After returning to Japan, in 1981 he set up SoftBank. The name was a portmanteau, 'soft' from software, but instead of the obvious financial connotations of 'bank', the second half of the word came from Idea Bank, the title of his notebook in which for many years he logged hundreds of potential inventions.

An early investment in the internet portal Yahoo set Son on the road to riches, but an exclusive deal to bring Yahoo to Japan was bigger. He was emboldened to believe that anything was possible.

Son was a rarity in the Japanese business world: charismatic, visionary and willing to take on the establishment. He and Soft-Bank brought the dynamism that more mature Japanese technology firms such as Panasonic, Toshiba, Sony and Fujitsu gradually lost. He was unfailingly polite, but of course he wasn't infallible.

Not all the companies he backed were winners. Around the same time as the Alibaba investment, SoftBank put money into Webvan, an overhyped grocery start-up that typified the era, only to lose a reported $160m. Son could spot trends, but he could also be domineering. His bold, crystal-ball-gazing assertions that manifested themselves in a '300-year vision' that was supposed to drive long-term investment decisions appeared to be rarely challenged by his coterie of advisers.

Still, he bestrode the globe in the company of fellow billionaires. Son invested in Japanese media alongside Rupert Murdoch and bought the US computer trade show Comdex from Las Vegas gambling tycoon Sheldon Adelson. Together with Microsoft co-founder Bill Gates, he implored the South Korean President Kim Dae-jung to throw all his resources into broadband investment to rebuild the nation's economy.

Looking down on the world from his 26th-floor office in Shiodome, a business district near Tokyo Bay, often wielding a bamboo sword to aid decision-making, Son showered money as he liked. Buying Vodafone's Japanese arm in 2006 was audacious; striking a deal for the business to distribute the iPhone exclusively from 2008 more so. In October 2012 SoftBank went one better, taking control of the lossmaking US telecoms company Sprint for $20bn, which at the time was the biggest US acquisition by a Japanese company.

If the Sprint deal did not announce Son's arrival in the upper echelons of the US technorati, then his next purchase certainly did. In November 2012, Son smashed the record price paid for a US family home when he acquired a mansion in Woodside, California, for a cool $117.5m.

The neoclassical-style property, set in a nine-acre hilltop estate, boasted a colonnaded pool house, tennis courts, manicured gardens, a detached library and panoramic views of the surrounding mountains. A short drive from the Stanford campus, Son had chosen to put down roots in one of the most exclusive neighbourhoods in Silicon Valley that counted Intel's Gordon Moore, venture capitalist John Doerr and the Oracle chief executive Larry Ellison among its residents.

## The Internet of Things

It was this sumptuous property that Simon Segars pulled up to on 27 June 2016, the latest in a series of meetings he had attended with Son over several years without ever really understanding their purpose. Some of them had taken place in tandem with Warren East, who had handed over Arm's leadership to Segars in July 2013 after 12 years at the helm, having expanded the company's range of licensees and, crucially, kept Intel at bay.

The 2004 acquisition of Artisan Components was one that some within Arm wished to forget. But in many respects it was the making of Segars, who uprooted his wife Rachel and their three children from Cambridge and relocated to the US in 2007 when it was clear the purchase was failing to gel.

'I said to Warren one day that the best thing I can do for the company was to go to California and try and fix this business because I think I understand it,' Segars said. 'He was very happy for me to go and do that.'

Artisan changed the culture of Arm, shifting its locus towards the US West Coast, where many of its customers were also based. The British firm even moved its US staff into Artisan's Sunnyvale offices. Assuming he succeeded in his mission to better integrate it and improve performance, the move made Segars favourite to become Arm's next chief executive long before there was a vacancy.

He had shown initiative to get there in the first place. Born in the town of Basildon in Essex to a fireman father and a teacher mother, engineering had attracted Segars from an early age: 'I have taken my fair share of things apart, soldered stuff, burnt my fingers and written some code on a computer,' he said in one interview.[4]

Standard Telephones and Cables (STC) was a big employer in the town, so Segars applied for a sponsorship and worked there for a year before departing for Sussex University, returning every summer vacation and after graduating with a degree in electronic engineering.

STC had a grand history, making telephone switchboards, undersea cables and early fibre optics for the nation. It was a major supplier to British Telecom and had benefited as the company upgraded its telephone exchanges following privatisation in 1984. But it suffered from a lack of international scale and its finances had been stretched by an unsuccessful foray into mainframe computers.

Opened in the 1960s, by the time the tall and angular Segars arrived at the dilapidated Basildon site it was a metaphor for its occupant, covered in scaffolding supposedly as a temporary fix to prevent concrete panels falling off.

What prompted Segars to hunt for another job was a developing interest in microprocessors and the stasis caused by the $2.6bn takeover of STC by its major shareholder, Northern Telecom (Nortel) of Canada. That deal was announced in November 1990, the same month that Acorn and Apple gave birth to Arm. Segars read about the launch in a trade magazine and was instantly curious. 'I wrote to the company, saying I was interested in designing microprocessors, please give me a job,' he said. 'They gave me an interview.'[5]

A few weeks after Arm's founders relocated to Harvey's Barn, Segars joined them as the 16th employee. He wasn't done learning. New colleagues pointed him towards Steve Furber at the University of Manchester, who agreed to oversee his master's degree in computer science. Segars' thesis on low-power processing was completed in two furiously busy years during which he also built the ARM7TDMI core that the breakthrough Nokia 6110 phone would use. Because Furber was still a regular visitor to Cambridge, at least Segars saved time from never actually going to Manchester while studying.

He progressed through senior roles in engineering, sales and business development, assuming more responsibility in April 2012 when a colleague, Mike Inglis, gave notice that he would leave to compete in a round-the-world sailing contest in 2013.

––––––––––

Just as Warren East had done, when he became chief executive Segars focused on finding new markets and new applications for the Arm architecture. Hence he was happy to meet big-spending business leaders such as Son who always had an eye to the future.

In the autumn preceding the Woodside dinner, Arm had outlined plans to increase its investment in new technology. On 15 September 2015, during a morning of presentations to City analysts and investors close to the Barbican arts centre in London, the company detailed how it would pour funds into areas such as networking infrastructure and computer servers to boost its presence.

Eye-catchingly, it boosted a target for market share in server chips from 20 per cent to 25 per cent by 2020, despite its negligible presence in that segment to date. Arm reckoned the addressable market would be worth $20bn by 2020, not far short of the forecast $25bn value of the market for mobile-applications processors that it dominated. But then Arm had to offer up something. The blistering growth in mobile devices that had largely sustained it since it joined the stock market 17 years earlier was starting to tail off.

Also discussed that day was the 'internet of things' (IOT), the next technological paradigm that would see low-power chips installed everywhere, from fridges to washing machines, so that appliances could be controlled and monitored from afar. Overall, Arm forecast the extra spending would grow to £40m in the 2017 financial year.[6]

Segars felt the constraints of the public markets keenly. Some shareholders supported the expansion of the cost base if it delivered long-term rewards; others berated him for shrinking margins. Internally, there was a perception that Arm could only invest so much. It was the balancing act of any publicly listed company. Still, it ploughed £215m into research and development that year, a healthy 28 per cent rise. At £108m, the cost of the annual dividend was rising nearly as fast.

The 2015 financial year was a strong one but the published numbers illustrated the constant need to spend more. Arm's profits were up 24 per cent to £512m, having more than doubled in four years, on revenues that rose by a healthy 15 per cent to $1.5bn. Beneath the headline numbers, technology royalties, over which the company had little control and were based on historic designs, boomed by 31 per cent. Revenues from licensing, a signpost of future prosperity, barely grew. The company said the pause came after a marked expansion in its customer base following the introduction of its ARMv8-A technology several years earlier.

The eighth version (v8) of Arm's architecture marked its entry into the 64-bit realm and would power its tilt at the computer server market with instruction sets that could handle more memory demanded by larger applications. Acknowledging that even the mobile phone would need significantly more memory in time, it was a long-term plan. Arm's architects began playing with ideas in 2005, kicked off a project in 2007, announced v8 in 2011 and the company deployed its first products in 2013.

What Arm referenced but didn't major on as it mapped out the future was IOT. It was an opportunity that didn't require huge investment just yet – and nor was it worth talking up gains that could be many years away. Under the jaunty headline 'fish are better with chips' in its annual report that year, it told the story of South Korean farmers who were growing healthier fish by installing Arm-based sensors in remote regions to keep track of water quality and nutrient levels. Writing in the chairman's overview, Chambers admitted that where IOT was concerned, 'the growth rate and size of the market are still uncertain'.

Yet it didn't stop crystal-ball gazers attempting to put a number on it. In 2015, the management consultant McKinsey forecast that linking the physical and digital worlds could generate up to $11.1tn a year in economic value by 2025 – equivalent to about 11 per cent of the world economy.[7] And as the small group sat down to eat at Son's house in June 2016, IOT was a hot topic of conversation. Segars, refreshed from a break the previous week in Jamaica, where he had avidly followed the fallout from the UK Brexit vote thousands of miles away, was touting for business.

As he talked about the huge volume of data that would be thrown off from billions of extra connected devices, Son and his long-time lieutenant, Ron Fisher, listened with interest. How to extract value from all that data with the application of artificial intelligence was a long-term business play of which they liked the look. IOT could be the next stage of the digital revolution, but it was too far off to interest pension funds that often fixated on the delivery of quarterly performance and regular dividends.

By this time Son and Segars knew each other reasonably well. Even before SoftBank entered the mobile arena with the acquisition of Vodafone Japan in March 2006, East had paid Son a visit in December 2005 with Segars in tow, both marvelling at the panoramic views the office afforded over Tokyo. They had talked about collaborating around TrustZone, Arm's service to protect computer code and data, but nothing really transpired.

---

It was the Vodafone acquisition that eventually brought Arm squarely onto Son's radar. Convinced that the internet's axis was shifting from PCs to mobile, he went hunting for the handset that would put web access at consumers' fingertips.

Apple co-founder Steve Jobs agreed that 'the time has come when we can make the ultimate mobile machine', reported Son. At its heart, the investor figured the machine would have to contain a central processing unit derived from Arm.[8] That conversation with Apple, whose history had already been entwined with Arm's for the best part of 20 years, would prompt another, surprising, chapter for the British company.

In the smartphone arena, as Apple and handset makers using Google's Android operating system slugged it out, both sides using Arm-based chips, Son identified the brains behind the devices that were preferred even after the sustained effort by the industry giant Intel to break into the market. That dominance fomented his interest, even as he pursued larger deals elsewhere. In 2012 SoftBank bought control of Sprint, the number-three US mobile carrier, in a $20bn deal.

'If I hadn't bought Sprint then I would have gone for Arm – I'm certain I could have got them for a lot less at that point in time,' Son said.[9]

The Arm contact did not let up. In December 2014, Segars and East – by now departed from the company – met Son for dinner at the Savoy in London. Making up the quartet was Nikesh Arora, a smooth Indian executive who had been lured from Google to SoftBank a few months earlier.

When the 2016 Woodside dinner took place, it came at a frenetically busy time for Son. On 21 June, SoftBank announced it was selling its majority stake in Supercell, the Finnish video-games maker best known for its fantasy title 'Clash of the Clans' for $7.3bn to Tencent of China. It had been less than three years since the company had first bought in but it said in a statement it was a deal that had been 'driven by its continued focus on disciplined capital allocation, including further deleveraging'.[10] It followed the announcement three weeks earlier that the group would offload shares worth at least $7.9bn in Alibaba, the first sale since it made the original investment in 2000.

It appeared that the years of dealmaking were catching up on Son. At the end of March 2016, SoftBank was sitting on a mountain of interest-bearing debt worth $107bn. Those giant borrowings had begun to weigh on the group's shares. The last thing most chief executives would do is countenance another giant acquisition. But Son was not most chief executives.

One man who investors were hoping would reduce risk in the portfolio and pay down debt was Arora, who had been elevated to become SoftBank's president and chief operating officer in June 2015. Under the headline 'SoftBank Readies for Version 2.0', Son called out tech companies that faced decline after 30 years due to 'evolving technologies, changing business models and overreliance on founders'. Arora, he said, 'will work with me to drive the transformation of SoftBank as it enters a new phase'.[11] He could not have been more clearly signposted as Son's heir apparent.

So Arora's exit, announced hours after the Supercell disposal, and on the eve of SoftBank's annual shareholders' meeting at which he was meant to be put forward for re-election to the board, came as a shock. Arora had not been universally liked and some investors questioned his supersized remuneration package. There was even some chatter about a conflict of interest with Silver Lake, a US private equity group, where he had an advisory role. But his departure was unsettling, nonetheless.

'I've been seriously hoping to pass on the baton at an early date to someone younger before I become a bottleneck,' Son explained the following morning to 2,200 SoftBank shareholders who had gathered in Hall A of the Tokyo International Forum for the group's annual meeting. He disclosed a previously secret plan to hand over the reins to Arora on his 60th birthday in August 2017. 'With a year or so to go . . . there are still things I want to make and I feel somewhat greedy,' Son added.[12]

'He works 18-hour days. He's invigorated. He said he wants to continue,' said Arora, whose exit reportedly stemmed from an impromptu discussion about the handover because it was mentioned in one of the AGM presentation slides.[13] Whether Arora shared Son's enthusiasm for pursuing Arm was unclear. There was speculation about a disagreement, but SoftBank said his resignation was not connected with the acquisition.

## Sold

AT 7.24am on 18 July 2016, the UK's business community awoke to a tweet from the Chancellor of the Exchequer, Philip Hammond. On only his sixth day in the job, the brand-new finance minister was cheerleading what for many felt like a devastating blow: the sale of Arm, the country's flagship technology company.

'Decision by SoftBank to invest in @ARMHoldings shows UK has lost none of its allure to global investors – Britain is open for business', a post from Hammond's Twitter account said. To underscore the point – and before shareholders, staff, customers and regulators had been given the opportunity to make up their own minds about the takeover – at 7.27am another tweet added: 'This would be largest ever Asian investment into the UK & would double size of Arm's UK workforce. Big vote of confidence in British business.'

The government stamp of approval had been carefully choreographed with a statement released to the London Stock Exchange just after 7am. The deep-pocketed SoftBank, known to few in the UK, had swooped to pay a cool £24bn in cash for Arm in a shock deal the Cambridge company's board was recommending its shareholders accept. In a note to the customers of the company's designs, the chief executive Simon Segars said it was 'one of the most significant days in Arm's history' but tried to reassure them that 'we remain in a business as usual mode'.[14]

It was a febrile time for the UK. The nation was still reeling from the aftermath of the 23 June Brexit referendum that determined it would depart the European Union economic bloc after 43 years as a leading member. In the days following that knife-edge verdict, an earthquake had engulfed British businesses and the political leadership who had expected the public would choose to remain. Old certainties crumbled: the pound fell sharply against the dollar and the Bank of England was forced to cut interest rates to steady investor nerves.

And now this. The UK had been a happy hunting ground for buyout houses and overseas acquirers, especially when debt was cheap. Airports, steelworks, telecoms providers, chemical firms and food companies with decades of history had all passed into foreign hands with little or no state intervention and barely a murmur of protest.

Cutting-edge ideas that originated from the UK's widely respected academic community went the same way. One example, DeepMind, an artificial-intelligence start-up co-founded by Demis Hassabis, a former child chess prodigy and neuroscientist, was bought for £400m by Google in 2014, just two years after it was set up. Autonomy, a FTSE 100 data-sorting company that was larger and longer established, was sold to Hewlett-Packard for $11bn in 2011 – although the deal later descended into acrimony.

Arm was different. It was a crown jewel, a technology champion, anchor tenant of Cambridge's world-leading knowledge hub and, many thought, unsellable. By creating an industry standard, just as its creators had dreamed of in 1990, Arm had put itself at the centre of the ultimate ecosystem. It was the 'Switzerland of semiconductors' that was trusted to do business with archrivals Apple, Samsung and Google simultaneously. Whenever takeover gossip did the rounds, it was easy to rebuff by sketching out the damaging response of, say, Google if Apple proved to be the purchaser. As Arm's first chief executive Sir Robin Saxby had replied to the casual inquiry from Intel more than 20 years earlier: don't buy the company, buy a licence instead.

The ownership of UK assets had been put into the spotlight just one week earlier. Theresa May launched her campaign to become the next prime minister with a speech delivered at a conference hall in Birmingham on 11 July in which she berated the previous government for almost allowing AstraZeneca, one of the UK's top drugmakers, to be sold to US giant Pfizer, 'with a track record of asset stripping and whose self-confessed attraction to the deal was to avoid tax'.

May added that: 'A proper industrial strategy wouldn't automatically stop the sale of British firms to foreign ones, but it should be capable of stepping in to defend a sector that is as important as pharmaceuticals is to Britain.'[15] Minutes after the speech ended, May's only rival, Andrea Leadsom, pulled out of the race, and she was driven back to London, preparing to take charge of the country.

Masayoshi Son was unperturbed by the tough talk or the latest political weather. Guided by his 300-year vision of the future, he flew into London the following day to secure his prize. By 18 July he was striding into Downing Street, grinning and shaking hands with Hammond on the steps, accompanied by Stuart Chambers, Arm's chairman.

Later that day, May stood up to speak in the House of Commons on a matter of national security. But first of all she welcomed the SoftBank acquisition as 'a clear demonstration that Britain is open for business, as attractive to international investment as ever'. And then she went on to the pressing matter of the renewal of the UK's nuclear deterrent, Trident, and the £31bn cost of four new submarines.

Son, late for his own press conference because of the ministerial glad-handing, talked about his concerned friends who were mulling whether to relocate their headquarters out of the UK after the Brexit verdict. 'I am totally opposite,' he told reporters effusively. 'I say this is the time that we invest with a strong commitment and belief in the future of the UK.'[16] And then he was driven to Cambridge to visit Arm's base for the first time and meet the senior team.

———————

Events had unfurled at breakneck speed. Only two days after their dinner at Son's mansion, and heading into the Independence

Day holiday, Segars received a call. 'I'm very excited,' Son said to Segars. 'I want to meet with you and your chairman this weekend.' There was no doubt that something was afoot.

Segars tracked down Chambers, a keen sailor, on board his 21-metre (70ft) cruising boat during a family holiday in the sparkling eastern Mediterranean. The idea of suggesting dates for the second half of July was quickly quashed.

Chambers had a layover planned in Marmaris, before sailing back west. Son flew into Dalaman airport 60 miles away with an armed escort and despatched another private jet to California so Segars could meet them there.

Chambers knew plenty about international takeovers. He had led the historic UK glassmaker Pilkington when it was acquired by Nippon Sheet Glass a decade earlier and even ran the enlarged business for six years until 2012, relocating to Tokyo for three of them. He had also chaired the FTSE 100 drinks-can maker Rexam, whose $6bn sale to Ball Corporation of Colorado had completed just a few days earlier.

Having spent his early life in Brunei, Malaysia, Singapore and Sri Lanka as his father travelled as an engineer for Shell, Chambers brought a global perspective to the boards he sat on. And his downtime was nearly as exciting as his corporate exploits, having scaled the peaks of Kilimanjaro and Fuji, sailed yachts across the Atlantic and once been held at gunpoint for four hours at Checkpoint Charlie before the Berlin Wall fell.

The lunch was convivial and Son, in short-sleeved shirt, cream slacks and deck shoes, was charming. Then he got to the point. 'I've been doing a lot of work,' he said, 'and I've been thinking I would like to buy Arm.' Son slid a piece of paper across the table containing the price he was willing to pay.

The seriousness of his intent was clear from the start. The 'internet of things' (IOT) was coming, and Arm had a big role to play. With SoftBank behind it, Arm could invest harder and faster than as a public company. And it wasn't just the money. Son sketched out his commitment to Cambridge and the UK, creating many more jobs.

Arm's board had a defence committee that met once a year to scan the horizon for such an event and to ask what else it should be doing so that it was not deflected from its profitable growth strategy. Son's interest in the company was known because he had talked about it for years. But no one thought he was in a position to act on it.

The action moved to London, where a phalanx of bankers and lawyers were mobilised on both sides. SoftBank's first three bids – including the one presented at the Marmaris lunch – were rejected, but Chambers and Son eventually shook hands at a meeting that took place in the upmarket Lanesborough hotel that faces out onto Hyde Park.

SoftBank's £24bn cash offer constituted £17 a share, a figure 43 per cent higher than the previous Friday's closing price, 69 per cent more than the average price over the previous three months and 41 per cent above Arm's all-time high reached that March. The Brexit effect had not made Arm any cheaper to buy, since it reported in dollars and, in a flight to quality, its stock had rallied 17 per cent since the referendum.

The board had been meeting every few days, attempting to measure the offers against what value Arm could deliver on its own. Directors judged that a point had been reached; the premium on offer was so high that most shareholders were unlikely to turn it down.

On Sunday 17 July, before he flew back to London, Son spoke to Theresa May to check for any obstacles before the announcement the following day.

---

The condemnation was quick. Hermann Hauser, whose foresight gave birth to the original Arm chip design, labelled it 'a very sad day for technology in Britain'.

Others could not avoid introspection at the news. For Cambridge Enterprise, part of the University of Cambridge that helped academics spin out their ideas commercially, Arm was 'the local kid who made good' and the head of a crop of unicorns created locally in the last 20 years. If only the majority weren't all getting bought by overseas investors.

'At the moment we simply are not able to take companies like Arm to the next level, to grow and retain our own Googles and AmGens,' wrote Tony Raven, Cambridge Enterprise's chief executive. 'Having become very good company-starters, we now must become finishers. Twenty years ago we set our sights on getting tech start-ups to the $1bn mark; now we need to reset our target at $100bn.'[17]

Other members of the old guard were more balanced. 'To me, in this internet of things idea, there are things that Arm is doing and things that SoftBank is doing and if you can put those two ideas together and make them better and bigger and move faster, it's a great win,' Sir Robin Saxby told the BBC. 'If you do it appallingly, it's the opposite of a win, right – which is a disaster.'[18]

There was also bafflement. On 18 May 2016, Arm had acquired Apical, an expert in computer vision technology, for $350m. The company, based in Loughborough in the English Midlands, would be useful in the IOT world because it enabled cameras to better understand their environment and adapt to weather changes. But until that point, Arm had very little in the way of IOT technology, insiders said.

Optimists countered that Arm had worked on IOT microcontrollers and explored the different ways to connect devices. Its critical contribution would be one of security: driverless cars talking to traffic lights risked lives if the communications system was hacked or faulty. All in all, IOT was a bet on a hard-to-imagine future, just as the 'portables' listed on Arm's first SWOT analysis in 1990 offered no clue of the mobile-phone boom to come.

Son pledged to put Arm at the 'center of the center' of Soft-Bank, boldly estimating that within 20 years one trillion Armchips would enter the market annually, a 67-fold increase compared to the 15 billion that had been shipped in 2015.[19] And because SoftBank didn't make microchips itself, it was well placed to maintain Arm's neutrality.

Calls of congratulation came into Son from the cream of the technology world: his old friend Jack Ma, Apple's Tim Cook and Paul Jacobs, the executive chairman of Qualcomm. 'From here, we can build relationships of trust with global leaders as partners and talk about how technology will develop five or 10 years down the line,' Son said.[20] It was exactly what Arm had already been doing.

In Japan, the news sent SoftBank shares plunging. All the signs were the funds it had raised from disposals were to pay down debt. But Son said later, 'We sold shares in Alibaba and Supercell that we didn't want to sell in order to come up with the money (for Arm).'[21] It had quietly begun buying shares in the company in April.

Back in the UK, national security concerns were slight. Segars and Chambers visited the Ministry of Defence to explain Arm's chip architecture, how its designs appeared in the Trident nuclear deterrent and why a sale to SoftBank made it no more likely to fall into Chinese hands than before.

Certain investors were a tougher crowd. Baillie Gifford, the Edinburgh fund manager that owned 10 per cent of Arm across a range of funds, criticised the takeover for being short-termist, although some of its holding was voted in favour of the deal when it could not muster enough support to block it.

'Arm had the chance to become a global technology giant,' James Anderson, the feted fund manager who ran Baillie Gifford's investment trust Scottish Mortgage and reaped rewards from his long-term backing of Elon Musk's electric car maker Tesla, wrote less than a year later. 'To do so it needed the willingness to invest at the cost of immediate profits, managers prepared to dream rather than calculate, and supportive shareholders to back this vision. We failed on all accounts. But who cares if the immediate uplift to performance is enough for a bonus?'[22]

UK political leaders were wary of false promises from overseas buyers ever since US food group Kraft bought chocolate maker Cadbury's in 2010 and soon after reneged on a commitment to keep open a plant in Somerset. Son's pledges over the Marmaris lunch to nurture Arm were hardened into post-offer undertakings.

SoftBank committed to keeping Arm's global headquarters in Cambridge and to at least double its 1,700-strong workforce so it could continue to develop leading-edge technology in its home market. It said it would also increase the company's non-UK staff, which made up 60 per cent of Arm's overall headcount of 4,200 people, over the following five years. And, so it could not be accused of stuffing the firm with low-value workers, on the fifth anniversary of the deal's completion, at least 70 per cent of both UK and non-UK staff must be classified as technical employees – a proportion broadly in line with Arm's historical trends. 'We won't be hiring janitors,' Alok Sama, SoftBank's chief financial officer, said.[23]

Its aftercare of the asset was well thought through but Soft-Bank's due diligence beforehand was light-touch to say the least. Over not more than two days its lawyers made some limited enquiries into how Arm structured its intellectual property contracts. Minor details such as these were not going to get in the way of Son's vision and the consummation of a putative 10-year love affair. Nor was the takeover contingent on gaining regulatory approval. If, for some reason, it was blocked, that was Son's risk to take.

In the end, Arm's sale was swept through at an unprecedented pace, supported by more than 95 per cent of its shareholders. Just 64 days elapsed from the Marmaris lunch to completion on 5 September, astonishing the bankers and lawyers that were involved.

By some token, nothing had changed. Arm would still maintain from Cambridge its design rulebook that determined how billions of devices operated. It would still be an essential node in a global, growing industry. And, if SoftBank made good on its promises, it would be expanded too.

And yet it was hard to shake off the feeling that something had been lost. Illogical as it seemed, because companies were bought, sold and reconstituted every day, the subtle power that Arm wielded felt somehow diluted because the UK's flagship technology firm had been subsumed by an overseas acquirer.

Time would tell whether that creeping unease would prove to be well founded. One thing was certain: at a time when sovereign power was beginning to understand the critical importance of the tools that governed future technology, it was a deal that would never have happened in numerous other jurisdictions.

## Chapter 11

# GOING GLOBAL: CHINA PLAYS ITS TRUMP CARD

## The US Defence Strategy

President Trump, the bolshy property developer and reality TV star, spent the early months of his surprise presidency wielding tax breaks to persuade big business to bring jobs and investment back to the United States from lower-cost destinations abroad. On 2 November 2017 it was the turn of the microchip giant Broadcom to put America first, to borrow one of Trump's campaign slogans.

'Let me say my mother would never have imagined that one day her son would be here in the Oval Office in the White House standing beside the President of the United States,' Hock Tan, Broadcom's diminutive leader, said proudly, dark hair slicked back and a stars and stripes badge pinned to his lapel.[1]

'And my mother too!' interjected President Trump, Tan's host that day, who darted forward to grasp the executive's shoulders playfully amid peals of laughter.

Since 2006, Tan had run Avago Technologies, which in 2016 became Broadcom after paying $37bn for the business of the same name based in Irvine, California. It was a transformative deal, creating a vast microchip empire whose output enabled data centres to function and smartphones to communicate.

Although it was domiciled in Singapore, Avago had a great American heritage. It traced its roots to the chip-making arm of Hewlett-Packard, one of Silicon Valley's founding firms. When it was carved out as an independent business by private equity investors Silver Lake and KKR in 2005, with backing from two Singaporean funds, Temasek and GIC, it chose the island nation for its corporate base even though most of the management would remain in San Jose, California. True, the tax terms were beneficial, but the business could argue it was local, too. HP had been making calculators in Singapore since 1973, right back when Tan was still an MIT student.

In choosing where it located itself, Avago could also see the benefit of being close to its customers. At inception, some 60 per cent of its revenues came from Asia and that proportion was only going to rise. Even the firm's new name was selected with growth markets in mind. Broadly, Avago can be translated into Chinese characters meaning 'peace of mind' and 'outstanding technology'.

All of those factors were cast aside by Tan as he stood in the Oval Office. 'Today we are announcing that we are making America home again,' he said, as Trump looked on gleefully. Broadcom, the employer of 7,500 people across 24 US states, declared that irrespective of tax reform its base in San Jose would become its sole headquarters. Its legal address would be in Delaware. Trump was fulsome in his praise, calling Broadcom 'one of the really great, great companies'.[2]

The president's 'America First' plan sought to rebalance the nation's global role, reduce its trade deficit, boost manufacturing and be less deferential to international bodies such as the World Trade Organization. Getting a company like Broadcom on side was extra sweet. As the microchip industry had grown, governments wielded relatively less spending power, and staking a claim on part of the complicated global supply chain was trickier as technology sped across borders. What hadn't changed since John F. Kennedy turned to Fairchild Semiconductor's integrated circuits to power the Apollo programme 50 years earlier was that technical advantage in the industry translated into power on the world stage.

Broadcom also had another motive. It hoped the timing of its re-domicile would smooth the path for its $5.9bn acquisition of Brocade Communications Systems, a specialist in storage area network switching kit, announced the previous November. Company takeovers by foreign investors, no matter how American they looked, risked scrutiny from the Committee on Foreign Investment in the United States (CFIUS), a panel that operated across government agencies to consider threats from corporate transactions to national security.

But rather than just focusing on Brocade, Tan was an inveterate dealmaker who had his next, far larger, target already in his sights. Just four days after his trip to the White House, he unveiled a blockbuster, unsolicited $130bn takeover of Qualcomm, a maker of next-generation wireless chips and a US corporate titan. The proposed combination would create the world's third-largest maker of microchips, behind Intel and Samsung, creating 'a global communications leader with an impressive portfolio of technologies and products', Tan affirmed in a statement.[3]

Whatever bonhomie had been manufactured for the cameras in the White House quickly ebbed. Qualcomm was a national and regional champion and had been rooted in San Diego, California, since its foundation in 1985 by Irwin Jacobs, a one-time professor of computer science and engineering at the city's university. The company had even put its name to the local sports stadium that had hosted Super Bowls and Major League Soccer. More importantly, Qualcomm (a portmanteau of Quality Communications) was a key developer of 5G technology vital for use in the defence industry, among others, after proliferating the code division multiple access (CDMA) technology standard for radio communications. And it had no intention of losing its independence.

With customers such as Apple and the Pentagon, the US government was attuned to the sensitivities of a change of ownership that might challenge its hegemony. On 29 January 2018, Qualcomm asked CFIUS to take a look. On 4 March, the review was widened and Qualcomm was ordered at short notice to delay its annual shareholders' meeting. One day later, a letter from the US Treasury to the two companies' lawyers raised potential national security concerns that had been identified by CFIUS. Much of the detail was classified, but the letter disclosed that these concerns related to Broadcom's 'relationships with third party foreign entities'.[4]

It went on to cite Broadcom's 'private equity' style approach to deals, which supposedly focused on short-term profitability at the expense of research and development (R&D) spending. 'According to press reports, in the last dozen years, Broadcom has spent six times as much on acquisition as on R&D, and former employees allege that it underinvests in long-term product development,' the letter read. This approach, if applied to Qualcomm, would weaken its competitiveness and influence and 'would leave an opening for China to expand its influence on the 5G standard-setting process'.[5] The charm offensive was forgotten; Tan's takeover was in deep trouble.

---

Where Japan was once viewed as a great technological threat to the US, and the Soviet Union its main military opponent, now China combined both concerns in one. The battle for strategic and technological primacy between the two superpowers ranged across the world. Unlike Japan, whose prowess owed much to

US patents to begin with, China was intent on developing a self-sufficiency that meant it could decouple from the US-dominated supply chain and sell its own technology, powered by home-grown microchips, internationally. It was undoubtedly home to the 'third party foreign entities' to which the US Treasury letter referred. Its potential had the US on red alert.

CFIUS had been set up by executive order in 1975 amid growing concern over OPEC countries buying American stocks and Treasury securities.[6] Its role was formalised by the Exon-Florio Amendment, which was added to the Defense Production Act in 1988. Exon-Florio, named after the two sponsoring politicians, granted the president authority to block transactions deemed harmful to national security, informed by investigations carried out by CFIUS.

The amendment's passage through Congress was propelled by worries over the proposed sale in 1987 of Fairchild Semiconductor to Fujitsu of Japan. At the time trade relations with Japan were rock-bottom, with several dumping suits filed in the memory-chip wars. Amid fears of American economic decline and increasing dependence on foreign suppliers, ceding a major supplier of computer chips for the military to a strategic rival won few fans. Fairchild's part in US microchip folklore – as the company of Noyce and Moore that begat the mighty Intel and numerous 'Fairchildren' – only served to heighten emotions.

Political and business opponents failed to persuade President Reagan to block the deal, but the French seller of the firm, Schlumberger, abandoned its plan after five months anyway, citing 'rising political controversy'.[7] Soon after, loss-making Fairchild was bought cheaply by US-based National Semiconductor. It didn't have much of a future, but at least it would be allowed to wither in American hands.

The Broadcom story played out differently to Fujitsu, but 30 years apart the result was the same. Last-ditch pledges from Broadcom to set up a $1.5bn fund to train the next generation of US engineers and not to sell any critical national security assets to foreign companies fell on deaf ears. On 13 March 2018, Trump issued an executive order to block the planned takeover, saying that it 'threatens to impair the national security of the US'.[8]

It was only the fifth time since CFIUS was created that a deal had been blocked by a sitting president based on the committee's advice. The last time was still fresh in the mind. On 13 September 2017, Trump refused the $1.3bn acquisition of Lattice Semiconductor by the Chinese investment firm Canyon Bridge Capital Partners on national security grounds. Lattice made field-programmable gate arrays (FPGAs) that allowed companies to put their own software on silicon chips for different uses, and it had previously supplied the US military. Canyon Bridge, the latest in a line of Chinese companies keen to buy US technology providers, was found to be funded partly by China's central government.

Unsuccessful bidders often get the chance to withdraw their bid before they are blocked, to avoid embarrassing publicity – or else they can just walk away, as Schlumberger did from Fujitsu when the heat became too much. However, the trend for blockage via executive order was accelerating. Of the five occasions it had happened, four had taken place since 2012. There were other trends too. Broadcom was the third out of these five contentious deals involving the microchip industry. But what particularly stood out was that this was the first time where the bidder was not Chinese – even though the perceived Chinese threat from a positive outcome was clearly driving decision-making.

There was no doubt that Broadcom's tactics were clumsy. Hock's move on Qualcomm so soon after his trip to the White

House looked arrogant. But the speed and scale of the US government's response was breathtaking.

Consider too that no takeover was agreed. And not only was Broadcom not a Chinese company, but its shares were also traded on the US exchange Nasdaq, its largest shareholders were familiar US institutions, it was run by a US citizen, it had significant operations in the US and, as of April that year, it would have once again become an American company, as celebrated so publicly by Trump.

That Broadcom would hobble Qualcomm and that China would as a result pull ahead in 5G were leaps of faith – but the US government could not take that risk. What was striking was that if Tan had waited five months to make his move, CFIUS would have had no leverage over it. Then, how odd it would have been for concerns over Chinese dominance to seep into a purely domestic transaction, a consolidation play in an industry that demanded huge financial outlay, arguably the creation of a new national champion. Yet so critical had the microchip industry become that the animal instincts of Wall Street to buy and build jarred with Capitol Hill's urge to protect.

Combatting China played to Trump's pugnacious economic nationalism. The threat it posed loomed large during Trump's successful election campaign. The country had also had a starring role in the latest US National Security Strategy. 'Every year, competitors such as China steal US intellectual property valued at hundreds of billions of dollars,' the December 2017 document said.[9] 'Stealing proprietary technology and early-stage ideas allows competitors to unfairly tap into the innovation of free societies.' Mistrust ran deep.

Perhaps unsurprisingly, CFIUS was given a broader remit soon after the Broadcom case. On 13 August 2018, the Foreign Investment Risk Review Modernization Act (FIRRMA) was signed into law. It gave CFIUS powers to scrutinise real-estate deals, minority investments and joint ventures into which technology assets were transferred. They were all paths that Chinese technology investors had gone down to try to access American works.

---

Concerns over China's efforts to accelerate its industry presence had also troubled the previous administration. Two weeks before Barack Obama left office in January 2017, the President's Council of Advisors on Science and Technology (PCAST) issued a report on protecting US leadership in semiconductors in response to concerns about China's state-led involvement in the sector.

The nation would not remain a market leader if it simply focused on making microchips more cheaply and easily, warned council members including Craig Mundie and Paul Otellini, respectively Microsoft and Intel senior alumni. It needed to innovate in many directions, such as exploring new ways of performing calculations and using materials other than silicon to build chips. Pointedly, PCAST also recommended the government 'reshape the application of national security tools, as appropriate, to deter and respond to Chinese policies'.[10]

Whether the Chinese really believed that Broadcom's cost-cutting ways would have allowed it to gain ground on the US in 5G is hard to determine. But in July 2018, it had an opportunity to retaliate.

At the same time as being Broadcom's prey, Qualcomm was also predator. In October 2016 it had announced a $47bn takeover of NXP, the former semiconductor arm of electronics group Philips, that would push it into a new growth area of automotive chips. Unlike the unwanted Broadcom approach, this one had the backing of NXP's board, but governmental approvals still needed to be obtained.

Nine jurisdictions had a say on Qualcomm, an American corporation and the world's biggest smartphone-chip maker, buying the Dutch NXP. Eight out of nine agreed, but China, which accounted for nearly two-thirds of Qualcomm's revenue in the previous year, failed to do so. Its State Administration for Market Regulation remained impassive. It did not expressly oppose the deal, but merely ran down the clock on the takeover without sharing a view, so that Qualcomm had no other option but to abandon its move.

The US Treasury Secretary, Steve Mnuchin, who incidentally chaired CFIUS, aired his disappointment in an interview on CNBC. 'We're just looking for US companies to be treated fairly,' he said. With glorious understatement, Steve Mollenkopf, the Qualcomm chief executive, observed: 'We obviously got caught up in something that was above us.'[11]

Then, on 1 December 2018, a glimmer of hope. After talks between Trump and the Chinese President, Xi Jinping, on the sidelines of the G20 summit in Buenos Aires, the White House said in a statement that China was 'open to approving the previously unapproved' deal for Qualcomm to acquire NXP 'should it again be presented'.[12] Qualcomm did not take up the invitation.

Still, this accommodation could have been seen as a sign of thawing relations between the two battling superpowers. But anyone who thought so was severely mistaken.

## China on the Offensive

Over several decades spent working in the microchip industry, Charles Kao had earned a reputation for getting things done. Slight and unassuming, with dark hair, a high forehead and typically dressed in an anonymous grey business suit, Kao specialised in the cut-throat world of memory chips, where plunging prices, multiplying capability and tight margins were a fact of life.

When the Formosa Group, one of Taiwan's foremost conglomerates, wanted to get into this market in 1994, Kao, a veteran of Intel and TSMC, was called in to help make it happen. To kick-start the production of 16-megabit DRAM chips – which could store 16 million digits of information – one of Formosa's divisions, Nanya Plastics, signed a licensing agreement with Japan's OKI Electric.

In its time, OKI had made telephone systems, minicomputers, printers and fax machines, but the deep recession of the early 1990s meant it was slimming down and needed to save money on manufacturing and development. In March 1995, a new business, Nanya Technology, was incorporated, a fab was built and a plan drawn up to push into 64-megabit DRAM chips by 1998. A deal to transfer technology in from IBM soon followed and, from a standing start, Nanya had momentum.[13]

In time, Kao – not to be confused with the Chinese-born, Taiwan-raised Nobel Prize-winning professor of the same name who pioneered the use of fibre optics in telecommunications – became president of Nanya Technology, which grew rapidly to become the world's fourth-largest maker of DRAM chips. He also chaired Inotera Memories, a manufacturing joint venture set up between Nanya and the German chipmaker Infineon in the same field. In a nation that applies titles to those it venerates, small wonder Kao earned the soubriquet of 'Godfather of Taiwan's DRAM industry'.

So when China wanted experts to help it make in-roads into the same market, it knew exactly where to come. On 6 October 2015, shock news broke that Kao was departing Nanya Technology. He wasn't just quitting for another Taiwanese firm, which would have been surprising enough. He was heading to Tsinghua Unigroup, the state-controlled, commercial spin-off from Beijing's highly rated Tsinghua University, which was investing heavily in new technologies.

Kao was the most senior hire yet in China's sustained raid on Taiwan's microchip talent. He would become Tsinghua Unigroup's executive vice president of global operations. Through gritted teeth, Nanya confirmed to local journalists that Kao had submitted his 'retirement application'.[14]

For a nation state for whom China had always loomed large across the Taiwan Strait, this latest incursion was akin to being under attack. Fed up with merely being the world's largest consumer of microchips, China wanted to make more of them too. It was a market in which Taiwan, a territory that China thought was rightfully theirs, excelled. In fact, about a quarter of microprocessors it used originated in Taiwan. If it wanted to catch up and eventually pull ahead, here was a good place to start. China had deep pockets, but it also needed intellectual property, reliable manufacturing and sufficient numbers of talented engineers.

'This will greatly hurt Taiwan's semiconductor memory industry,' Allen Chu, the chairman of Marbo Investment and a regular media commentator on business matters, told one local news bulletin about the Kao defection. 'It will be impossible for us to stop future poaching of staff.'[15]

---

China set out its ambition in 2014. Analysts say the immediate prompt was most likely revelations from the whistleblower Edward Snowden of US cyber-intelligence efforts that made clear to China how much relying on American-made communications equipment posed a national security risk.

In response, the country targeted an annual growth rate of 20 per cent up to 2020, by when it would be a volume producer of 16nm and 14nm chips, aided by potential financial support from the government of up to 1 trillion yuan, equivalent to $170bn. There was also a clearer focus on developing national champions and segment winners and allowing private equity firms to allocate funds.[16]

In May 2015, China put more meat on the bones of its plan. The Made in China 2025 strategy embraced all high-end manufacturing but had microchips at its heart. The country's aim was to produce domestically 40 per cent of all core components it needed by 2020, lifting to 70 per cent by 2030. For chips, it outlined a target range of 41–49 per cent by 2020 and 49–75 per cent by 2030, which would make it as a nation the number-one chip manufacturer worldwide. Compared to its chip output of less than 20 per cent in 2019, it was a huge statement of intent. If it was successful, the impact would be to cut in half imports of US semiconductors in 10 years and eliminate them entirely from the Chinese supply chain within 20 years.[17]

From the outset, it looked like mission impossible. As the management consultancy McKinsey pointed out in a 2015 research note, if Chinese manufacturers were to hit the 2025 self-sufficiency goals the government had laid out, roughly all the extra foundry capacity installed globally over the next decade would have to be in China.[18]

And it had tried and failed to build up its semiconductor industry before, often for cultural not economic reasons. Chinese companies were fast followers, not great innovators. As the factory to the world they could churn out products quickly and cheaply, but not always to the highest standard. That was fine for consumer electronics but not for microchips, where precision was everything.

In 1990, the Microchip Project 908 devoted 2bn yuan to closing the gap between Chinese microchips and those produced internationally. Bureaucracy meant it took seven years for production to begin, a lifetime in chips. While it was eager to build up its civilian capabilities to mirror its military prowess, the party saw little return for its money.

The state-run system did not always support innovation. Far from creating champions, it created competitors that duplicated each other and tended to spread money thinly. Without region-alisation there was no cluster effect as seen in Silicon Valley or similar specialist hubs. Commercial companies were reluctant to invest. When China had been allowed to buy companies, it strug-gled to integrate them. And it had a well-known cavalier approach to intellectual property that made foreign investors tread carefully.

---

When China set out its stall in chips, Taiwan Semiconductor Manufacturing Company (TSMC) was riding high. In 2014, it commanded 54 per cent of the foundry segment for global semiconductors, with 42 per cent of revenues originating from manufacturing processes of 28nm and below.

Taiwan took great pride in what it had built, labelling the company its 'sacred mountain'. Morris Chang, the frustrated executive who was lured from the US to Taiwan almost two decades before, was now the elder statesman of industry, regu-larly representing Taiwan on the global stage. TSMC had built the vast economies of scale required for a capital-intensive industry and a standard process technology that it constantly sought to improve. It was a strong and indispensable incumbent in the foundry industry, rather like Microsoft had been in the market for computer operating systems.

And it had seen off allcomers. The Common Platform Alliance (CPA) was a powerful collaboration between IBM, Samsung and GlobalFoundries, the chip manufacturer that traced its roots back to AMD. Together they intended to develop a new foundry busi-ness model, but failed to make a dent in TSMC over the course of a decade and eventually threw in the towel. In another sign of the old guard peeling off, in October 2014 GlobalFoundries received $1.5bn to take IBM's loss-making chip-manufacturing unit off its hands.

Samsung, the third member of the trio, had also suffered a blow that for TSMC was a giant vote of confidence. The Taiwanese firm was called in to manufacture on behalf of Apple, perhaps the most exacting client in the world, later easing out Samsung.

The Apple relationship kicked off over dinner, when Chang hosted the company chief operating officer Jeff Williams for a simple home-cooked meal. Apple first contracted TSMC to make all the application processor chips for its iPhone 4 devices in 2010. Chang reportedly committed $9bn and a 6,000-strong workforce at its Tainan site to show that he took the opportunity very seriously.[19]

Not everything it did had worked, though. TSMC's own effort to develop its own DRAM memory chips without relying on foreign technology was long forgotten. But the memory market's cyclicality, often caused by factory over-capacity, made margins extremely volatile compared to the more stable non-memory segment where it excelled.

---

TSMC had kept at bay Semiconductor Manufacturing International Corporation (SMIC), one of China's notable early industry successes. Founded in 2000 by Richard Chang, a Taiwanese expatriate with years of experience from Texas Instruments and TSMC (and not thought to be related to Morris), its creation was not without bad blood. Chang took 180 staff with him from TSMC when he set up in Shanghai.

Suspicions arose that it was stealing more than just TSMC's people, because SMIC was ready for business astonishingly fast. A US lawsuit lodged in 2003 was designed to make US customers take fright. Allegations of furtive photocopying and email exchanges led to an eventual settlement where SMIC would pay $175m to TSMC over six years. But a second lawsuit was filed when SMIC reportedly did not hand back stolen documents and carried on compiling more information. Eventually, six years after the first filing, SMIC agreed to pay TSMC $200m and grant it shares and options for a 10 per cent stake in the company. Chang left soon after the case was resolved in 2009.[20]

Made in China 2025 was better organised than previous efforts. A $20bn national integrated circuit fund was incorporated to support new ventures and local government funds sprang up too. There was easy access to equity and cheap debt.

What China had on its side this time was that its home market had changed dramatically. Rather than simply requiring a microchip supply to manufacture export goods, China had a vast consumer market too. By 2005, China was the world's largest consumer of semiconductors; by 2012, it was purchasing more than half of world output. In the decade leading up to 2014, China's semiconductor consumption grew at a compound annual growth rate of 16.7 per cent compared to 4.7 per cent for the worldwide market, according to a PwC study.[21]

If only it could get the product right. Its own microchip industry accounted for 13.4 per cent of the market, PwC found. The chips travelling in each direction were markedly different. Top-of-the-range chips were arriving to be installed in smartphones and consumer electronics for distribution internationally. It still produced lower-end chips.

Regardless, China was prepared to throw money at the problem by investing in its capabilities. It would also attempt to buy its way to dominance. Soon after setting bold targets, in July 2015 an informal $23bn bid by Tsinghua Unigroup for Micron, the last US memory-chip maker of any scale, was knocked back. The company, which a generation earlier had led complaints against

Korean imports that were being sold at less than fair value, wasn't interested. And in any case, the deal would surely have fallen foul of CFIUS.

But as well as trying to buy assets it needed people – and ideas. History repeated itself when Richard Chang resurfaced in 2018 as the leader of SiEn (QingDao) Integrated Circuits Company, which was hiring in Taiwan for staff willing to relocate to the north-eastern port city of Qingdao. It reportedly began testing a 300mm wafer production line in August 2021.

Chinese firms went hunting for staff in South Korea and Japan but had most luck in Taiwan, where culture and language were most alike. Many Taiwanese engineers found the lure hard to resist. Chinese employers dangled salaries double or triple what they were currently being paid, plus regular free trips home, sub-sidised accommodation and private education for their children. Recruits were encouraged to bring their experience, train local staff – and were often offered greater incentives if they committed for five years or more.

Despite becoming a hub for world-class engineering, Taiwan had somehow kept salaries low for its highly trained technicians and hoped loyalty and national pride counted for something. Fearful of losing key staff – and their valuable secrets – Taiwanese firms warned workers to be careful. In summer 2018, the Reuters news agency photographed in the reception of a chip company in Hsinchu, Taiwan, a leaflet that featured a blonde-haired woman with a finger pressed to her lips. 'Protect Company Competitive Edge', it read, 'No Confidentiality in Public'.

The ideas might have had some protection, but the manpower still slipped. In December 2019, Nikkei Asia reported that Taiwan had lost 3,000 engineers to positions at mainland companies – nearly a tenth of the sector's estimated 40,000-strong workforce.[22]

There was always suspicion about what new hires were bringing with them. The US got wind of an egregious example, Micron, which remained a company of great interest to China even if it could not be bought.

Micron had close links with Nanya Technology. In 2008 it bought out Infineon, now called Qimonda, from Inotera Memories, which by this time was operating two 300mm wafer fabrication facilities in Taiwan, producing a total of 120,000 wafers per month. Six months earlier, Nanya and Micron had struck a DRAM memory-chip joint venture and pledged to develop and share future technology. By 2013, Micron had negotiated exclusive access to all of Inotera's manufacturing output and by 2016 it had bought out its partner entirely.

On 10 October 2020, UMC, Taiwan's second largest foundry business, pleaded guilty to stealing secrets from Micron on behalf of China's Fujian Jinhua Integrated Circuit Company, one of three companies China hoped would become a DRAM leader. UMC hired several engineers from Micron's Taiwan operation – the rebranded Inotera – who passed confidential information to Fujian as it raced to begin production of memory chips at its new $6bn facility.

The court heard that as the police raided UMC's offices, a junior employee was passed incriminating USB drives, laptops and documents, which she stashed in her locker. The company agreed to pay a $60m fine.

'UMC stole the trade secrets of an American leader in computer memory to enable China to achieve a strategic priority: self-sufficiency in computer memory production without spending its own time or money to earn it,' the US deputy attorney general Jeffrey Rosen said.[23]

In the end, the subterfuge was in vain. Fujian Jinhua went out of business when the US imposed export restrictions on it.

China would have to try different tactics – or hope that different companies had more luck. One thing was for sure – it would not stop in its efforts to lead the microchip industry.

## A Dutch Lesson

One place from where China could draw lessons for its micro-chip mission was the Netherlands. Just like Arm in Cambridge, the home of ASML in Veldhoven, a small southerly town close to Eindhoven, was far removed from the industry's twin axes of America and Asia. And yet in the space of several decades, it had become an indispensable equipment supplier to anyone aspiring to manufacture the most advanced semiconductors.

Its lithography machines projected a chip's intricate blueprint design onto a silicon wafer, which was then developed like a photographic film with chemicals that etched into place the pattern of conductors and insulators required to connect and isolate millions of transistors and other components. And then it was repeated, on the same wafer, up to 100 times.

Building and transporting these machines was every bit as complicated as the process they carried out. The size of a small bus, each machine comprised 100,000 precision components and two kilometres of cabling that from Veldhoven were transported in 20 blue-and-white trucks the 80 miles to Amsterdam's Schiphol Airport before being gently loaded through the nose of three 747s.

Since spring 2017, ASML had been selling its latest model, the NXE:3400B, which promised manufacturing gains, higher-power products and performance enhancement by making a reality of a long-held dream: commercial microchip production at the 7nm node. Even with a sticker price of €160m, the company could not keep up with demand. It had planned to ship 20 units in 2018 and expand production capacity to at least 30 systems in 2019.

Incredibly, more than five decades after Moore's Law was conceived, it was predicted that this machine would preserve it for at least another 15 years. For chipmakers, that made it worth the wait and wild expense. The technology that came to market in the NXE:3400B had taken more than 20 years to develop at a cost of more than €10bn.

A description of how it worked read like a passage from a science fiction novel. It began with ASML's extreme ultraviolet (EUV) technology that used light with a wavelength of just 13.5nm to shrink the size and increase the density of a microchip's features. That wavelength was 14 times shorter than ASML's previous best, known as deep ultraviolet (DUV), which used light of 193nm.

To create its own EUV stream, ASML devised a system that heated droplets of molten tin with a carbon dioxide laser to a temperature 100 times higher than the surface of the sun. Some 50,000 droplets were generated per second; each must be blasted twice by the laser in quick succession.

From this exercise came plasma, a gas of ions and free electrons, that emitted EUV radiation which was bounced between six of the world's smoothest mirrors – these mirrors, if enlarged to the size of Germany, would not have a bump higher than a millimetre. For its showstopping finale, the beam hit the silicon wafer with the precision equivalent to shooting an arrow from Earth to hit an apple placed on the moon.

ASML – which stood for Advanced Semiconductor Materials Lithography until it was no longer spelled out – had not got this far on its own. The imperative for world-class expertise plus financial constraint made collaboration essential. In fact, the company didn't make most of the machine's parts. Many came from a family of some 600 suppliers, of which 60 were judged to be key, including optics company Zeiss, based in the southern

German town of Oberkochen, which produced the six specially shaped mirrors costing €1.5m each.

---

Creating a world leader was not fast work. In 1984, the same year that Steve Jobs proudly unveiled his Macintosh computer and Steve Furber and Roger Wilson pondered what would become the first Arm chip design, ASML began life as a joint venture between two Dutch companies, Advanced Semiconductor Materials International (ASMI) and Philips.

ASMI had already made a name for itself as Europe's first major supplier of chip-making equipment to Intel, Motorola and others, including kit that deposited thin chemical films on the silicon wafer during the manufacturing process. Philips was an electronics giant that traced its roots back to 1891 and its activities were eclectic. The company was famous for making light bulbs, electric razors and, from 1982, compact discs. Philips had developed a so-called 'stepper' – the shorthand name for a step-and-repeat camera that evolved into a lithography machine – for internal use at its microchip factory in Nijmegen near the German border, at the time Europe's largest facility of its kind.

Just like the rationale for Acorn spinning out Arm to win Apple's backing, Philips recognised that supplying the wider industry would only work if its steppers were placed into an arm's-length company that was easier to do business with. It assigned 50 scientists to the task, later setting up the group in an unprepossessing wooden shed with a hole in the roof on its site in Eindhoven.

The fearsome cost of development meant ASMI did not stick around for long. That also explained why the market quickly thinned from 10 lithography tool suppliers at ASML's inception. Coming back from the brink of bankruptcy several times, the company found itself in a three-way battle for market leadership with two Japanese firms better known for their expertise in cameras, Canon and Nikon. The PAS 5500, a machine that met the expectations of the microchip colossus IBM, helped it push into second place in the early 1990s.

———

All the while, EUV was being dreamed about by chipmakers who had already seen the industry advance from using mercury vapour lamps as their light source and now wanted to build transistors smaller than could be designed with the wavelength of visible light. Understanding that it was key to extending the life of his eponymous law, Gordon Moore called it 'soft' X-ray.

EUV's germination was truly international, from Russian mirror research in the 1970s, to the first images projected in the research department of Japan's Nippon Telegraph and Telephone Corporation (NTT) in the mid-1980s. Despite great scepticism, laboratories supported by the US Department of Energy (DOE) took up the baton. To access their ideas, in 1997 Intel formed a consortium, EUV LLC, that included peers Motorola and AMD. They struck a cooperation agreement with the DOE to license its technology, putting in $250m over three years to cover development, including equipment and researchers' salaries. Without the cash, the development of EUV is likely to have petered out.

Efforts to broaden the consortium's knowhow by bringing in overseas partners sparked a row in the US Congress over giving foreign corporations access to sensitive intellectual property. With recollections of the memory-chip wars still fresh, Nikon of Japan was deemed an unacceptable joiner. The US did not have the same troubled trading history with the Netherlands, so an ASML partnership was approved in 1999 – although it took more than a year to negotiate.

Niggles surfaced in 2000, however, when ASML spent $1.6bn to acquire Silicon Valley Group (SVG). The deal mopped up what remained of the lagging US lithography industry and was an elegant solution that bought ASML's way onto Intel's supplier list. (Although its progress meant that in all likelihood it would have eventually got there anyway.)

The other Japanese competitor in this field, Canon, had been blocked from forging an alliance with SVG years before. This time, fears that selling SVG would risk transferring its technology into enemy hands took on cartoonish proportions in the early months of George W. Bush's presidency. The US Defense Department held up the deal's approval when it emerged that ASML's chairman Henk Bodt sat on the board of Delft Instruments, a Dutch company that nine years earlier had been fined for illegally shipping night-vision goggles to Iraq. In some quarters, it didn't seem to matter that the episode took place long before Bodt joined Delft.

'We had a hell of a task to battle through and we had to give on various things,' says Doug Dunn, the straight-talking Yorkshireman who became ASML's chief executive in 2000. 'I have never been exposed to that level of geopolitical influence and rhetoric but we never opted to walk away because we recognised that over the years the gain would be tremendous.'

Dunn thought the acquisition would take three months to complete. By the time ASML agreed to sell SVG's Tinsley Labs division, which had supplied mirrors to the Hubble space telescope, and committed to invest heavily in the remaining US operation he had made several trips to the Pentagon and a year had passed. Nevertheless, ASML's position was strengthened considerably and soon after, in 2002, its time-saving Twinscan innovation, a machine which could handle two silicon wafers at once, saw it crowned as the world's top lithography tool supplier.

In 2006, it demonstrated it was nosing further ahead. ASML delivered two EUV prototype models, to the Interuniversity Microelectronics Centre (IMEC) in Leuven, Belgium, and the College of Nanoscale Science and Engineering (CNSE) of the State University of New York. Both institutions were tasked with conducting extensive research into the new technology system. By now, it was clear it worked, but expert help was needed to see if it could be made to work commercially. Additional financial help was required too. In 2012, ASML's expectant customers – Intel, TSMC and Samsung – ploughed billions of dollars into the business in a final push to get EUV over the line.

ASML's story demonstrated that even in markets riven with fierce competition, sometimes it required broad collaboration to make advances. 'Everyone works with everyone else, until you go commercial and then you fight to the death. That's how the game is played,' said Frits van Hout, who was ASML's executive vice president and chief strategy officer until April 2021. For all the masculine posturing of the microchip industry, no company – and no country – could push the boundaries on its own.

## A Downward Spiral

ASML's story could not be further from that of Huawei, the Chinese telecoms supplier that found itself at the centre of an increasingly nasty trade war centred on technology and microchips.

On 1 December 2018, while changing planes at Vancouver International Airport, Meng Wanzhou, Huawei's chief financial officer, was arrested at the request of the US, who wanted to extradite her to face fraud charges, and held in a detention centre.

Meng had come to view Canada as a second home. For a time, she had been a permanent Canadian resident and some of her four children were educated there.

As international recriminations mounted, Meng was bailed, put under house arrest in her six-bedroom mansion on the corner of a leafy Vancouver suburb and issued with an ankle bracelet so her movements could be tracked. At a court hearing she was charged with financial fraud and deceiving four banks to evade US sanctions against Iran. Ten days later, China detained two Canadians, Michael Kovrig and Michael Spavor, alleging they were spies. Meng countersued the Canadian government.

The diplomatic incident took on extra importance because Meng was also the daughter of Huawei's founder and chairman, Ren Zhengfei. Incidentally, her arrest took place on the same day that Presidents Trump and Xi met at the G20 summit in Buenos Aires.

It was not hard to see why Huawei became a lightning rod for that battle. It operated across several interlocking segments – networking equipment, microchips, mobile-phone handsets – and its influence was growing. Unlike most Chinese companies, it had also made a good fist of internationalising, and in at least one field – advanced networking equipment – where the US had no comparable operator. Huawei was employee-owned too, so perhaps an easier target than a state-owned firm. With its technical excellence and deep defence industry knowledge, its progress was troubling to see for the home of the innovative Bell Labs and the Massachusetts Institute of Technology (MIT). And President Xi wanted more firms just like it.

———

Huawei was set up in 1987 by Ren, an engineer in the People's Liberation Army for nine years until 1983. It was founded with a few thousand dollars after he lost his job amid a sweeping round of military cuts. Ren was also a member of the Chinese Communist Party (CCP).

For the western media that toured the vast campus in Shenzhen, southern China, the highlight was trying to catch a glimpse of the company's own version of the White House, which was perhaps symbolic of the market it would most like to crack. 'The colour is yellow, not white,' Ren insisted in one interview.[24]

Huawei had spread across the US like knotweed, taking share from its two largest rivals in telecoms base stations and other infrastructure, Ericsson and Nokia, the Nordic neighbours who had taken an early lead in mobile handsets but since retrenched. As 4G technology was deployed, the Rural Wireless Association said it thought a quarter of its members used kit from Chinese suppliers such as Huawei and ZTE, another provider.[25]

Huawei's detractors said it was little more than an arm of the CCP and owed its breakneck growth to sharp hardware and software subsidies. Never mind the profit, those critics insisted Huawei installed technology at a loss so it could grab sensitive data through concealed 'backdoors'. The company denied spying, although the US argued it could be pressed to do so by China's 2017 National Intelligence Law. Regardless, if the US didn't act soon, military experts envisaged a scenario where sensitive American communications would be carried over Chinese-built networks.

Critics had a suspicion that its infrastructure expansion was a warm-up for something even more insidious. For 2018, Huawei's sales soared 20 per cent to 721.2bn yuan ($107.13bn), with the consumer business the largest contributor. Huawei had set a goal in 2016 to be the world's number-one smartphone maker by 2021 and, even though it had limited distribution in the US, now just Samsung stood in its way.[26]

Powering its rise in handsets was HiSilicon, Huawei's fabless chipmaking division that was only established in 2004 but had grown to reportedly employ 7,000. Its Kirin chipset added to US concerns because it was regarded as competitive with the best that Apple and Qualcomm had to offer. HiSilicon was thought to be the largest designer of integrated circuits in China, producing chips based on the Arm architecture for smartphones, fixed and wireless networks, including the coming 5G standard. Not only was it winning market share, the company was also showing signs of self-reliance. Huawei disclosed that 8 per cent of the 50,000 5G base stations it sold in 2019 came with no US technology whatsoever.

It was not the sort of development that would warm frosty international relations. The US and China had briefly rubbed along well and trade relations normalised as China joined the World Trade Organization (WTO) in 2001. That was a marked improvement from when the US supported the Chinese nationalists against invading Japanese forces during the Second World War. They became the exiled Republic of China government in Taipei. Finding themselves on opposite sides in the Korean War (1950–3) guaranteed several decades of limited US relations with mainland China. Matters only picked up after US President Richard Nixon met Chairman Mao on a trip east in 1972.

The more recent downward spiral can be blamed on trade. In 2010, China became the world's second-largest economy and five years later passed Canada to become the largest US trading partner. It was not an equal relationship. In that year, 2015, the US trade deficit with China was $367bn, with almost half due to the purchase of computers and electronic parts. As the gap widened, recriminations flew.

In the aftermath of publicising its Made in China 2025 plan, President Xi Jinping had increasingly mused in public about how to correct what he viewed as China's technological deficiency, while Trump, outspoken on the presidential campaign trail, focused on what he said China had taken from the US already – and how to take it back.

In a speech at China's first Work Conference for Cybersecurity and Informatization, held on 19 April 2016, Xi said the fact that the internet's core technology was controlled by others was 'our greatest hidden danger'. No matter how big a Chinese internet company became, if it relied on the outside world for key components, 'this can be compared to building a house on another person's foundation, however large or beautiful it is, it might not stand the wind or the rain, or might even collapse at the first blow'.

Core technology, he added, was 'a national treasure' that required 'indigenous innovation, self-reliance and self-strengthening', but that did not mean China should 'close the doors to do research and development, we must absolutely persist in open innovation' to test itself against the best, Xi proclaimed.[27]

Trump was more direct. Two months later, on 28 June 2016, he set out his trade policy during a campaign rally at an aluminium recycling centre in Monessen, Pennsylvania. In what used to be America's steelmaking heartland, Trump said China's entrance into the WTO 'has enabled the greatest job theft in the history of our country' and, as secretary of state, Hillary Clinton had stood idly by 'while China cheated on its currency, added another trillion dollars to the trade deficit, and stole hundreds of billions of dollars in our intellectual property'.[28]

Stood before a wall of scrap-metal bales, Trump pledged to 'use every lawful presidential power to remedy trade disputes'. He invoked Ronald Reagan's 100 per cent tariff on semiconductor imports, introduced in 1987, 'that had a big impact, folks. A big impact.'[29]

Such rhetoric explained why, in the weeks after Trump was elected US President in November 2016, he took a call from Tsai Ing-wen, the President of Taiwan, offering her congratulations.

As the first contact between a leader of Taiwan and an incumbent or incoming US President in nearly four decades, it was a conversation loaded with meaning. 'Interesting how the US sells Taiwan billions of dollars of military equipment but I should not accept a congratulatory call,' Trump later tweeted.

The US trod a fine line with China and Taiwan. Under its 'One China' policy, Washington acknowledged Beijing's position that there was only one Chinese government, and it did not have formal diplomatic relations with Taiwan. China did its best to freeze out Taiwan from international organisations.

But since the Cold War, it was obvious the island's location made it strategically important for the US and its allies in Asia. It maintained unofficial contact and 'strategic ambiguity'. And through the Taiwan Relations Act of 1979, the US supplied weapons and pledged support when it needed to defend itself, essentially in the event of an attack from China.

What changed over the years was the build-up of China's military might. From the US perspective, the fall of Taiwan would be an unacceptable sign of the changing world order. The island's strength in microchips that lent it great power in a modern, digital economy only upped the ante. In 2016, Taiwan was the 22nd-largest economy in the world and a key trading partner to the US.

On 18 October 2017, President Xi declared at the opening of the China Communist Party's five-yearly national congress: 'We have sufficient abilities to thwart any form of Taiwan independence attempts.' He called for talks between the two sides as long as Taiwan recognised the 1992 consensus, an alleged agreement of the existence of 'One China'.

———

Against that backdrop, a trade war was already simmering. On 7 March 2016 the US Department of Commerce had put Chinese telecoms equipment maker ZTE onto its 'entity list', meaning US companies needed a licence to sell it goods or services. The list was first published in February 1997 to track organisations with links to weapons of mass destruction but was expanded to take in national security and foreign policy interests.

ZTE was found to have broken US sanctions by selling to Iran and North Korea. For failing to settle a $1.2bn penalty, a denial order was imposed in April 2018, which meant ZTE was banned from buying US technology components. However, Trump's intervention meant the order was lifted in July that year after a $1bn financial penalty was paid. It was an unpopular move with some in his own political party. The ZTE incident turned out to be a rehearsal for the main event.

Two weeks before Huawei's Meng was detained, the US proposed criteria to identify 'emerging and foundational technologies' that were essential to US national security and should be subject to future export controls under the new Export Control Reform Act. And after Meng had been put under house arrest, in late January 2019, the US Department of Justice accused Huawei of financial fraud, money laundering, conspiracy to defraud the US, obstruction of justice and sanctions violations. Ren, the company's founder, hit back, saying it was unfortunate that the US regarded 5G as a strategic weapon. 'For them, it's kind of a nuclear bomb,' he said.[30]

On 15 May 2019, the US upped the ante. It banned American companies from selling goods or services to Huawei without first obtaining a licence, which in practice was a blanket ban that meant it could no longer use Google's Android operating system for its smartphones, for example.

Although it was not an American company, the British chip designer Arm instructed employees to halt 'all active contracts, support entitlements, and any pending engagements' with Huawei and its subsidiaries. In a company memo, it said its designs contained 'US origin technology'.[31]

That missive proved to be premature. Like a number of other technology suppliers, it had to establish the 'nationality' of some of its products and, bizarrely, in some instances that came down to analysing timesheets to determine which staff worked on which product where.

The result was that Neoverse, Arm's group of processor designs for use in data centres and high-performance computing, as well as some of its Cortex varieties, which were all developed in Austin, Texas, were no longer sold to Huawei and support was stopped. However, the eighth version of Arm's architecture (v8) and subsequent ninth version (v9) that was launched in 2021 were still made available to Huawei's HiSilicon division because they were developed in Cambridge in the UK.

Effective on 19 August 2019, the US added Huawei affiliates to the entity list so that the company could not buy from American suppliers through subsidiaries in the UK, France or elsewhere. And, to mark the anniversary of the buying ban, in May 2020 the US tightened the noose further by amending its foreign-produced direct-product rule. That meant Huawei could no longer buy from foreign companies if they used American technology in semiconductor manufacturing and prompted TSMC to curtail supply from September. Its Kirin chipset was dead – for now.

The US did all it could to extend its influence beyond its borders. It convinced the UK to ban mobile providers from buying new Huawei 5G kit after the end of the year and remove it from their networks by 2027, even though it would delay 5G rollout by two to three years and add £2bn to costs.

And it was persuasive when its rules did not apply. ASML's prized EUV lithography machines did not exceed the 25-per-cent-by-value threshold of US-made components that meant they required a licence to ship to China. In 2018, ASML had been granted a licence by the Dutch government to sell its most advanced machine to a Chinese customer, but that summer the Dutch prime minister Mark Rutte was reportedly handed an intelligence report on the potential repercussions from the sale while on a visit to the White House. The US also had recourse to the Wassenaar Arrangement, a successor to the Cold War club that co-ordinated between 42 participating states export restrictions of so-called 'dual-use' technology that may have military as well as commercial uses. Soon after the report was handed over, the Dutch licence was revoked and ASML's delivery, thought to have been to SMIC, was not made.[32]

The hope of Eric Meurice, ASML's fund-raising chief executive who had retired in 2013, to make its technology available 'to every semiconductor manufacturer with no restrictions' proved to be wishful thinking. In December 2020, SMIC was put on the lengthening entity list too, one of 77 mainly Chinese additions, because it 'perfectly illustrates the risks of China's leverage of US technology to support its military modernization', Commerce Secretary Wilbur Ross said.[33]

---

As with the political intervention in the acquisition strategies of Broadcom and Qualcomm, many corporations did not agree with what was going on. In fact, they were furious.

Two months after the May 2020 directive, the electronics design and manufacturing trade body SEMI calculated that its members had already lost $17m of sales to firms unrelated to Huawei. 'The new restrictions will also fuel a perception that the supply of US technology is unreliable and lead non-US customers to call for the design-out of US technology,' it said. 'Meanwhile, these actions further incentivize efforts to supplant these

US technologies.'[34] Fewer sales meant less cash for research and development and less innovation, which would harm national security in the long run.

'The problem is that the US becomes the vendor of last resort,' said Wally Rhines, TI's former semiconductor chief, who aired fears over the prolonged use of the entity list. 'If you're in Japan or Europe, why would you use US semiconductor designs when you can never be sure your end-product can be sold to the Chinese or to anyone else that the US decides is an enemy?'

Rhines had history on his side. As chair of a technical advisory committee to the Commerce Department, he recalled how the 1979 Export Administration Act, which confined sales to US allies in the Cold War years, contributed to reducing the global market share of domestic semiconductor manufacturing equipment firms from 76 per cent to 45 per cent in a decade. Much of those sales were lost to Japan, which was also an ally but operated more liberally. The US imposed licence requirements on spare parts and user manuals.

US firms were no different to any other: the Chinese market had become too important to ignore. In 2014, Intel had invested $1.5bn in a subsidiary of Tsinghua Unigroup, which owned RDA and Spreadtrum, two of China's largest fabless design companies, as part of an accord to increase the adoption of Intel designs in the nation's mobile devices. In the same year, Qualcomm had backed a series of semiconductor start-ups and partnered with SMIC on 28nm products and 14nm process-technology development.

———

The tension reverberated beyond the US–China axis and sparked further instances of bilateralism. Japan was no longer a top-tier chipmaker but it still played a vital role in the industry's supply chain. On 1 July 2019 it halted preferential treatment on the export to South Korea of three ingredients vital for the semiconductor industry: hydrogen fluoride, a gas employed during the etching process; photoresists, used to transfer circuit patterns onto the wafer; and fluorinated polyimide for smartphone displays.

Japan reportedly produced about 90 per cent of the world's fluorinated polyimide and photoresists. By requiring exporters to gain permission each time they wanted to ship, a process that took 90 days, chipmakers had to scramble for alternative supply and were said to be days from running out.[35]

The spat was traced to a South Korean court ruling that ordered Nippon Steel to compensate former forced labourers and harked back to Japan's colonisation of the Korean peninsula before 1945. But the reaction was determinedly modern and threatened to cripple high-tech production. It set off efforts to reduce Korean reliance on Japanese supplies and had broader, cultural impact. Visitors to South Korea that summer saw shops proudly display signs in their windows declaring 'no Japanese goods'.

The high-tech world, which had advanced so far thanks to decades of collaboration, was splintering. But it was not an even split. When faced with a straight choice, most customers knew the Chinese alternative wasn't good enough – for now.

# PART THREE

ARM (2017-21)

## Chapter 12

# BIG DATA AND WALKING WITH GIANTS

## Too Much Energy

The 26 vacant hectares to the north-west of Dublin that had been earmarked to become Amazon's largest campus of data centres in Ireland were mired in controversy. Plans submitted by the internet giant in March 2017 were frustrated by protestors concerned about the environmental impact of the development, which would begin with one data centre measuring nearly 21,000 square metres (223,000 square feet), three times the size of the pitch at the Aviva Stadium, host to the city's raucous international rugby and football fixtures. Amazon had another seven centres in mind alongside this first one on the greenfield site. By the time contractors finally arrived to start work in autumn 2019 they were many months late.

Amazon was no stranger to Dublin. It already had one data centre nearby, one in the pipeline near the city's airport and three in Tallaght to the south. They were in good company. More giant, anonymous boxes run by digital behemoths including Google, Microsoft and Facebook were clustered around the edge of the city too.

It was a sign of the times. The mobile revolution had made portable a trillion tiny everyday transactions and the data they generated had to be stored somewhere. And just like the handsets they were continuously connected with, the data centres needed microchips to make them run. It was another market that Intel dominated – and another that Arm coveted. The pair were poised to re-run their rivalry all over again.

Inside the data centres, aisle after aisle of computer servers were studded with lights and humming softly. These facilities were the physical manifestation of the digital economy, where vast troves of information were stored, created from every click and swipe. Here was the computing 'cloud' or, to mix metaphors, the backbone of the internet, remotely serving up data for email, social media, banking transactions and online shopping at the touch of a button.

Ireland had long provided a convenient European base for a host of American technology companies. Just down the road was one of Dublin's corporate flagships, IBM's 100-acre campus, which had been opened in 1997. The country offered a good skills base, a common language, and the powerful pull of a low corporate tax rate. And as many industry players switched from offering manufacturing to services, its pitch to become a major data-centre hub had been astonishingly successful. What was surprising was the storm that these giant sheds provoked.

Amazon's new site was one of 10 data centres under construction at that time, adding to the 54 already in operation. Some 31 more had planning permission.[1] Ireland's cool climate offset some of the awesome heat that large volumes of computer processing generated, and it was ideally located to connect the US with mainland Europe, via transatlantic cables and Dublin's underground fibre ring. A recent study by Ireland's Industrial

Development Agency (IDA) found that the sector had contributed €7.1bn to the economy since 2010.

But that investment – and the modest number of jobs that came with it – had a cost. Data centres sucked vast amounts of power from the local energy grid to run them and keep them cool. Locals feared outages and blamed data centres for Ireland's poor environmental record. It was on track to miss its 2020 target for reducing carbon emissions and heading for a European Union fine.

Ireland's Central Statistics Office (CSO) suggested that electricity consumption by the country's data centres increased by 144 per cent between 2015 and 2020 – and rose to 849 gigawatt hours in the fourth quarter of 2020.[2]

An engineer, Allan Daly, one of the most visible complainants against Amazon's expansion, called for a further assessment of the site's energy demands before planning permission was granted. Daly's environmental concerns had already contributed to Apple's struggles a two-hour drive west.

In February 2015 the iPhone company announced plans to construct its first two data centres in Europe, including one at a forested site in Athenry, a small town in County Galway, where Daly lived. Planning was approved by September that year but the project became mired in protests and appeals.

The path was eventually cleared and, in November 2017, after a trip to see Apple chief executive Tim Cook at the company's headquarters in Cupertino, California, Ireland's Taoiseach Leo Varadkar said 'we will do anything that is within our power to facilitate' the centre getting built.[3] But Apple had lost interest and in October 2019, as workmen busied themselves on behalf of Amazon, it signalled the Athenry site would be sold.

Around the time that ground was broken in north-west Dublin, Ireland's electricity grid operator EirGrid published a sobering forecast. It said that the nation's total electricity demand over the next ten years would grow by between 25 per cent and 47 per cent, largely driven by new large users, many of which were data centres. EirGrid's analysis suggested that by 2028 data centres and other large energy users could account for around 29 per cent of all electricity demand.[4]

Ireland's challenges were not unique. In 2019, Singapore, which had become the data hub for Southeast Asia, imposed a moratorium on the building of new data centres. The Netherlands also paused construction. When the ban was lifted, the Amsterdam metropolitan area introduced a power budget and efficiency target for new developments. To get approval, each new facility required a power-usage effectiveness (PUE) ratio of 1.2, which was determined by dividing the amount of power run into a data centre by the power used to run the computing equipment it contains.

It was a case of different sector, similar problem. Two decades on from the breakthrough in mobile telephony, when power-efficient chips shrank battery sizes that enabled the introduction of lightweight handsets, Arm's designs were closing in on a new opportunity after years of trying.

Data-centre operators could promise the earth by building vertically, recycling waste heat and installing their own renewable energy capacity to supplement the grid. But the unassailable fact was that data usage was going up. If the internet giants didn't want to become neighbourhood pariahs, they needed to do everything they could to bring their energy consumption under control.

## Amazon Chooses Arm

On the chilly morning of 3 December 2019, Andy Jassy strode onto the stage at the Venetian Expo Center in Las Vegas. It was the second day of the re:Invent conference for which 65,000 technologists had dashed from their Thanksgiving weekends to attend.

Jassy had been Jeff Bezos's technical assistant, joining Amazon straight from Harvard Business School in 1997 a few weeks before the company went public on Wall Street. He was a key lieutenant in the diversification away from book sales and learned early on how exacting his boss, Amazon's laser-focused founder, could be. 'I thought that I had very high standards before I started that job,' Jassy said in one interview. 'And then . . . I realised that my standards weren't high enough.'[5]

Back in time, Amazon had been frustrated by how long it took internal IT projects to come to fruition, so it developed a common computing platform that could be shared across the business. Realising there could be an external market for its knowhow, Jassy gathered a team of 57 people in 2003 to work on an offering. From 2006, Amazon was renting out computing services including data storage to customers who did not want the cost and hassle of investing in their own hardware and were drawn by the flexibility of scaling usage up and down to meet demand.

So-called cloud computing was on the rise but for a long time Amazon Web Services (AWS) was as unassuming as its leader. Jassy was a fan of chicken wings which created a hand-made 'wheel of fortune' to spin and select who would present their ideas as a method of getting through bulging meeting agendas.

He was named the division's chief executive in 2016, by which time it was too big to ignore. In the 2019 calendar year, AWS reported sales of $35bn, a 37 per cent rise – almost twice the rate at which the entire Amazon empire was growing and marking a tenfold sales increase in just six years. At $9.2bn, its financial contribution was almost two-thirds of the group's operating income for the period.[6]

It wasn't just big within Amazon. AWS commanded a 45 per cent market share of the computing cloud, streets ahead of its closest competitor, Microsoft's Azure platform, which accounted for 17 per cent according to market watchers at Gartner.[7] Clients ranged across all sectors: Siemens, Halliburton, Goldman Sachs, Pfizer, Apple; and in the public sector, even the CIA.

---

Such was the interest in the business, it was unsurprising that Jassy was starting to take the spotlight. At the re:Invent conference, inaugurated in 2012 to excite AWS customers and partners about the road ahead for its technology, Jassy paced the stage for almost three hours, dressed in dark suit, blue shirt and white-rimmed shoes that would have not been out of place in a nightclub. And the highlight of the session for many was the unveiling of a new microchip.

Announced a year earlier, the Graviton chip was AWS's first custom-made processor to power its 'instances' – the industry jargon for the virtual servers that ran workloads for its clients in the cloud. Promising to deliver up to 45 per cent cost savings in some areas, early customers included the genealogy website Ancestry.com and South Korea's LG Electronics. IT partners such as Symantec and Red Hat had begun developing services to dovetail with the chip.

Now AWS was going further. In fact, a second version of Graviton was being worked on even as the team built the first. To a smattering of whoops and applause, Jassy unveiled Graviton2, reeling off its particulars. Compared to the original, it had four times more compute cores – an indicator of processing power – five times faster memory and seven times better performance. Arguably most importantly, he added, Graviton2 had a 40 per cent better price-performance than the latest generation of x86 processors from Intel. Before moving on to the next item, Jassy paused to exclaim: 'That's unbelievable if you think about that.'[8]

During its 25-year history, Amazon had ranged from books to clothing and groceries to become the 'everything store'. It had dived fearlessly into new markets, making its own movies, building electronic devices including the Kindle, data centres, and now microchips.

This latest diversification was part of a broader trend. The world's biggest technology firms recognised that computing power and performance had become a great differentiator for their business. Just like the political struggle along the US–China axis, they craved a leadership position and less reliance on others.

Awash with cash and already a huge buyer of chips, Amazon could afford to develop its own, tailored to its needs, reduce purchasing costs and throw down the gauntlet to the rest of the industry. Alongside it, Apple, Google and Facebook – the giant consumer-led internet platforms whose software and social media had long since overshadowed the original kings of Silicon Valley – were making a grab for the stuff that gave their home its name. This was the era of 'custom' silicon – when microchips became too important to leave to the microchip industry alone.

Customisation was nothing new. It harked back to the earliest days of the industry but had been ditched in favour of general-purpose chips with the advent of Intel's 4004 in 1971 that could be programmed to tackle different tasks. But the technology had moved on and tech giants had sufficient scale to make customisation extremely worthwhile.

Amazon had set hares running in January 2015 when it acquired an Israeli chip designer, Annapurna Labs, for a reported sum of up to $370m. It was a modest amount for the internet giant, but a statement of intent. Annapurna was founded four years earlier by Avigdor Willenz, who had previously set up another chip designer, Galileo Technologies, that was bought by Marvell in 2001 for almost $3bn.

Jassy said the deal was a 'big turning point' for AWS. It had worked with Annapurna on the first version of the AWS Nitro System, the foundation technology upon which its instances were built. Now it had the in-house skills to fashion its own chip.

The development was bad news for Intel, which had succeeded in transferring its dominance of the PC market into servers (where it had resumed battle with AMD) and on which AWS had relied in its formative years. While Intel remained a 'very close partner', and AWS was also being supplied by Intel's great rival AMD, Jassy pointed out that 'if we want to push the price-performance envelope for you it meant that we had to do some innovating ourselves'.[9]

The deal was far better news for Arm, on whose designs Annapurna's chips were based. As early as 2008 it had targeted the server market, knowing that the eighth version of its architecture, launched in 2011 and the first to embrace 64-bit processing, would pave the way. It forecast huge growth and hoped that environmental concerns over energy consumption and its own low-power credentials would aid progress. It made sense that the efficiency that prevented mobile phones from running out of juice prematurely or overheating batteries could benefit servers that gobbled electricity.

In its 2015 annual report, Arm cited the example of a fraud-detection system created by PayPal using its server chip designs. It had been half the price to buy, one-seventh of the cost to run and one-tenth the size of traditional data-centre kit.[10] But despite those factors, efforts to win over the industry had been dogged by false starts.

A start-up chipmaker, Calxeda, had been involved with HP on a project to use Arm server chips as long ago as 2011, but it went out of business. AMD's attempt to use Arm to replicate the success it had enjoyed with the Opteron chip a decade earlier was short-lived. Broadcom dropped its Arm-focused Vulcan programme for networking, storage and security applications when Hock Tan's Avago Technologies acquired the business in 2016. And Qualcomm abruptly shut down its Centriq division that produced Arm server chips in 2018. It appeared that an industry-wide attempt to break Intel's dominance had foundered.

The 2015 boast from Arm that it could grab a 25 per cent market share in servers by 2020 retreated into the middle distance. By the end of 2018 it had been pushed out to 2028, while Arm's actual share languished at just 4 per cent. It needed to make a breakthrough or call it a day.

So desperate was it, Arm even risked compromising its own prized independence by seeding the market. It invested in several firms that were developing server chips based on its instruction set, including Calxeda. Annapurna was another, where Arm veteran Tudor Brown was installed on the advisory board. Arm also struck a 'strategic partnership' with Marvell, which had acquired Cavium, a firm that bought the blueprints from Broadcom's Vulcan project to push forward its work.

Another investment that had shown great promise was Ampere Computing, a start-up founded by former Intel president Renée James, who worked under chief executive Paul Otellini for many years. Ampere was part-owned by the private equity group Carlyle and counted Microsoft and TikTok's Chinese parent company ByteDance among its customers. It picked up the X-Gene processor project that another firm, AppliedMicro, had worked on since 2011 and, in April 2022, filed for an initial public offering. Arm argued that this crop of investments, if they paid off, benefited the entire 'ecosystem' by leading to more software developers working on its server designs.

Although Amazon always kept its options open, the enthusiastic launch of Graviton2 demonstrated the market leader was throwing its weight behind Arm. The vote of confidence had parallels with Nokia's choice of TI's Arm-powered chip two decades earlier.

The market size might be smaller, but Arm could charge more per licence and per royalty. After years of hype, the company hoped the rest of the industry would finally sit up and take notice. And its green credentials might be a help too. AWS claimed Graviton2 delivered up to 3.5 times better performance per watt of energy use compared to its other processors.[11]

## Apple Takes Control

Apple's own journey into custom silicon began several years before Amazon. On 23 April 2008, it spent $278m to buy a promising firm, PA Semi, that employed around 150 chip engineers. Originally called Palo Alto Semiconductor, it had been set up five years earlier by Dan Dobberpuhl, a man whose superstar status in the industry was belied only by his modesty. At Digital Equipment Corporation (DEC), Dobberpuhl had been the lead designer on chips that powered its VAX minicomputers, and then the Alpha and StrongARM microprocessors that followed in the early 1990s. When Intel took ownership of StrongARM in 1997, he did not hang around. A subsequent venture, SiByte, which designed a 64-bit processor for high-performance networking, was next to benefit from his encyclopaedic knowledge, before being acquired by Broadcom. This time around, Dobberpuhl was wooed at Steve Jobs' home during a short courtship.[12]

The firms already knew each other well. When Apple switched its Macs to Intel chips, there had been speculation that its engineers had spent a long time working with PA Semi to see if its low-power chips were a viable alternative.

What was beyond doubt was that PA Semi represented a crucial move for Apple, setting it on the path towards silicon sustainability, squaring it off against industry leaders Intel and Qualcomm on which it relied. What might not have been clear at the time was that it was also a first step in disentangling itself from Samsung, a key supplier that would become an archrival soon enough.

'We have a historic relationship with that team and some of the experts there cut their teeth on doing Arm implementation,' said Warren East, Arm's chief executive at the time, when quizzed on the PA Semi acquisition on a quarterly earnings call the week after it sparked a flurry of interest. 'So, I think it's no bad thing if a little bit more Arm expertise finds its way into companies like Apple,' he added.

'But as to what they intend to do with it, we've no comment and you need to talk to Apple about that, because it's nothing to do with us.'[13] Two weeks before sealing the PA Semi deal, Apple had placed the call with East to ask for an Arm architecture licence.

---

'People who are really serious about software should make their own hardware,' the Xerox veteran and Dynabook's architect, Alan Kay, once said. Jobs referenced the comment when he launched the iPhone in 2007. As he stood on stage that day, showered in plaudits, he knew that the iPhone was good, but not as good as it might have been if Intel chips had been fit for its purpose and if Apple, with the help of Samsung, hadn't been forced to cobble together a chip from something originally designed for a DVD player.

Jobs had always been fascinated by the design aesthetic, enthusing in his 2005 Stanford commencement address about the 'wonderful typography' that his computers featured after he was inspired by calligraphy classes. Jobs put the customer experience first and worked backwards to the technology required to deliver it.

It was better for his customers not to worry about what went on inside. Former Apple chief executive John Sculley noted that the Macintosh was sold as a 'closed box' that required a special tool to open the casing – and if owners did so Apple would void the product warranty. 'Steve insisted on the restriction because he believed that software, not hardware, would become more important in the computer industry,' he wrote.[14] The iPhone was similarly hard to crack open; later editions came with a tamper-resistant screw.

It was all part of the Apple magic. But it was clear that Jobs obsessed over the minor details so customers didn't have to. The decision to take more control over its chips could have been sparked by the iPhone compromise. It could have come from a failure to agree with Intel over the use of its Atom chip in the iPad. It could have stemmed from the knowledge that PC makers had suffered at the hands of their chip suppliers, predominantly Intel, which grabbed the lion's share of sector profits, and mobile-device makers should not face the same fate.

Or it could have been that Apple wanted to try again, this time on its own. PowerPC, an alliance forged with IBM and Motorola to produce chips for its Macs, started well but was disbanded after failing to make sufficient progress. Announcing the Mac's switch to Intel in 2005, Jobs said bluntly, 'we can envision some amazing products we want to build for you and we don't know how to build them with the future PowerPC roadmap'.[15]

---

Now Apple had more money to fund a push, with cash, cash equivalents and short-term investments of $15.4bn on its balance sheet at the end of the September 2007 financial year. And just like Amazon would become in servers, it was a vast consumer of chips, with 52 million iPods sold in the most recent year. That scale offered the chance to create a virtuous circle: as long as

Apple's gadgets remained highly sought after, it could spread the steep development cost over big production runs. And the better the software it wrote, the fewer processor clock cycles it needed, the lower the power it consumed.

'Steve came to the conclusion that the only way for Apple to really differentiate and deliver something truly unique and truly great, you have to own your own silicon,' said Johny Srouji in a 2016 interview. 'You have to control and own it.'[16]

Srouji understood only too well, having been hired to mastermind the whole effort. The stocky Israeli microchip engineer joined Apple in March 2008, weeks before the PA Semi transaction, as a senior director for handheld chips and VLSI (very-large-scale integration), according to his LinkedIn profile. Born in the northern port city of Haifa, Srouji was a product of the Technion Israel Institute of Technology, the educational facility that had done much to make the country a hotbed of microchip talent. He studied undergraduate and master's degrees in computer science, before joining IBM and switching to Intel in 1993, where he worked on methods for testing how well semiconductor designs would work.

At Apple, the first fruit of his labour was the A4 chip. It was announced as part of the iPad launch on 27 January 2010: a new device, but also the first to contain a chip that Apple had designed itself, having taken out an Arm architecture licence.

A4 offered 'exceptional processor and graphics performance along with long battery life of up to 10 hours', the company said, but its debut gained little attention on the day.[17] The iPad's responsiveness was 'because of the custom silicon that we designed for this product', said Apple's Bob Mansfield, Apple's senior vice president, in one of the supporting promotional videos, adding that what the A4 chip provided was 'a level of performance that you can't achieve any other way'.[18] It featured again in June that year, powering the new iPhone 4.

Apple made a bigger splash with its chipmaking efforts in 2013 with the A7 chip for the iPhone 5s model, the first smartphone chip with a 64-bit core. The A7 enabled more features on the phone, including Apple Pay and Touch ID. It also put clear blue water between Apple and the competition for a year while providers that still used 32-bit cores raced to catch up.

———————

By this time, the chip design division had grown. After buying PA Semi, in 2010 Apple added Intrinsity, an Austin, Texas-based specialist in Arm design, for a reported $121m.

In a rush to get the A4 chip to market, Intrinsity was suspected to have provided its CPU core, effectively the chip's brain. Under contract to Samsung, Apple's chip partner, Intrinsity had designed an Arm-based processor called Hummingbird, a version of which analysts speculated had become the A4. For technical reasons it's an Apple-designed chip because Apple is thought to have released to Samsung the 'netlist file', essentially an abstract version of the final make-up.

Apple was intent on taking more control, not least because tighter integration between its processors and its operating system gave it an advantage over Google's Android operating system, which was powering many of its smartphone competitors.

That meant owning the graphic design system (GDS), the final file that is sent to the manufacturer with details of the chip's layout. It didn't matter that Samsung had prioritised Apple over other customers and given it access to the latest process node that delivered the smallest, most powerful chips. For the A6 chip, released in 2012, Apple solely handled design and Samsung's role was limited to manufacturing. And speculation mounted that Samsung wouldn't even be doing that for much longer.

———————

A blockbuster lawsuit launched in 2011 signalled the strain on their relationship, even though the divisions within Samsung that made smartphones and components including microchips operated at arm's-length from each other – and didn't always get on.

If Apple wanted to hurt Samsung, then shifting away the $5.7bn it had spent in the previous year on flash storage chips, RAM and microprocessors for its iPhones would have been a start. Including display panels, that figure was predicted to rise to $7.8bn in 2011.

But chip battles were never straightforward. The company, which was Samsung's second-biggest customer after Sony, knew there was no immediate alternative supplier for the quality and volume it required. Besides, this suit didn't come down to money: both companies were earning billions from the mobile revolution and were likely to continue to do so. It was a continuation of Jobs' 'thermonuclear war' against Android handset makers – and, by extension, Google, that had exploded a year earlier when it sued the Taiwanese outfit HTC.[19]

Samsung was one of the few phone makers that survived and then thrived when the iPhone's launch laid waste to the rest of the industry. Its Galaxy smartphone, which debuted in June 2009, was followed by the Galaxy Tab, a 175mm (seven-inch) tablet computer unveiled to the public in September 2010, eight months after Apple's iPad was first shown off.

Apple accused Samsung of 'slavishly' copying its devices. The suit included 10 charges of patent infringement related to the rectangular design of the iPhone and iPad and the use of gestures to operate the touch-screen. In retaliation, Samsung said its own patents had been infringed too.

The legal battle was protracted and fanned out in courtrooms around the world. By December 2015, when Samsung agreed to pay Apple $548m in settlement, but appealed another part of the ruling, their partnership was drifting further apart.

A year earlier, after Morris Chang and Apple's chief operating officer Jeff Williams had forged a relationship over dinner, TSMC was first called on to manufacture for Apple.

Regardless, when Apple and Samsung finally settled their differences out of court in 2018, after a jury in the US District Court of Northern California ordered Samsung to pay Apple $539m, both sides had prospered. Despite losing some of Apple's business, Samsung surpassed Intel as the world's biggest chipmaker for the first time in 2017 when its semiconductor revenues hit $70bn compared to Intel's $63bn. Once again, the tectonic plates of the industry had shifted.

---

On 11 October 2018, Apple went shopping again. In what was termed a licensing deal, it spent $600m to buy a chunk of the German firm Dialog Semiconductor, whose power-management chips had been used in iPhones for the preceding decade. Apple was buying patents, several Dialog offices and a team of around 300 engineers, most of whom already worked on chips for Apple. Half of the sum was a pre-payment for three more years of chip supply.

In the decade since the A4 chip, Apple had become a fully fledged 'fabless' semiconductor manufacturer. It produced its own chips to handle cameras, AI, and for its watches, TVs and headphones. A 2019 McKinsey study found Apple to be the world's third-largest fabless player, behind Broadcom and Qualcomm, and estimated that if it sold chips, its annual revenues from them would be up to $20bn and its semiconductor business would be worth up to $80bn.[20]

In December 2020, Qualcomm's shares slipped when Srouji told his staff that Apple was also working on its own cellular modems, the chips that connected its devices to the internet.

Srouji described the move as 'another key strategic transition' that was 'making sure we have a rich pipeline of innovative technologies for our future'.[21]

That disclosure was notable because 5G iPhones used Qualcomm parts and Apple had relied on the chip giant for cellular modems in the past – although the relationship had grown increasingly bad-tempered. In 2017, Apple sued Qualcomm for allegedly overcharging, and Qualcomm countersued, claiming Apple owed it money and had infringed some of its patents. At one point, Apple switched to Intel for its modem chip supply.

Peace broke out in April 2019 when the iPhone maker agreed to pay $4.5bn to Qualcomm and the pair struck a licensing arrangement that would run until at least 2025. Three months later, Apple demonstrated it would not put itself in the same situation in the future. It acquired Intel's smartphone modem unit for $1bn when Chipzilla gave up on the business line. That deal took Apple's portfolio to more than 17,000 wireless technology patents, ranging from protocols for cellular standards to modem architecture and operation.

In March 2021, news of Apple's €1bn investment in Munich, Germany, partly funding a silicon design centre focused on connectivity and wireless technologies, was further evidence of its desire to pull away from Qualcomm, just as it had from Intel and Samsung. People that knew Apple well estimate that Srouji, who was elevated to Tim Cook's management team in late 2015 and showered with stock options, had at least 4,000 engineers working under him. Apple set up satellite engineering offices close to the headquarters of Qualcomm and Broadcom and hired freely.

Analysts think 2023 is the year it will reveal its own, in-house designed, modem, giving it total control of the guts of the iPhone, whose chips are now made using the 5nm processor technology. Connection speeds, which were already important, have become vital, given the rise in augmented-reality applications that run on top of real-world sound and images. That trend, plus the complexity of wireless chips, has driven up their price to rival processor costs. By taking them in-house, Apple is banking on running these components more effectively together – squeezing an extra 10 per cent of performance or 15 per cent from battery life – and making them at lower cost.

It is a hugely expensive undertaking, but Apple's cash and equivalents swelled to $172.6bn at the end of the September 2021 financial year. Countries might not have been able to afford silicon self-sufficiency – but the world's largest companies were giving it a shot.

———————

Apple and Amazon were not alone in investing heavily in custom silicon. Google's Pixel 6 phone was powered by a homegrown microchip named Tensor, made for it by Samsung. In a blog post on 19 October 2021, Monika Gupta, a senior director at Google Silicon, set out new capabilities along a familiar theme. Google Tensor, she wrote, was capable of 'running more advanced, state-of-the-art ML (machine learning) models but at lower power consumption compared to previous Pixel phones'.[22] She gave the example of using automatic speech recognition on long-running applications such as a recorder without draining the battery quickly.

Chinese handset maker Huawei had long used its own Kirin chips, made by its subsidiary HiSilicon. And Alibaba, the Chinese trading giant, had developed its own server chips at T-Head Semiconductor, a wholly owned division founded in 2018. Facebook owner Meta was reportedly doing similar.

In 2019, Tesla got in on the act, showing off a custom chip that would run the self-driving software for its electric vehicles. 'How could it be that Tesla, who has never designed a chip before, would design the best chip in the world?' the company's founder, Elon Musk, asked modestly at the launch presentation. 'But that is objectively what has occurred. Not best by a small margin, best by a huge margin!'[23]

Even Samsung, once again the world's largest chipmaker in 2021 according to Counterpoint Research, is planning to make chips that it won't supply to anyone else.[24] New custom-made processors for its Galaxy range, due in 2025, can only come from forcing its previously federated semiconductor and smartphone divisions to work more closely together. For its part, Apple still relies on Samsung as a major supplier of the latest OLED (organic light-emitting diodes) screen displays.

Billions of dollars have been expended on the race to be different. But these microchip designers, whether long-standing players or newcomers to the field – including Apple, Amazon, Samsung, Qualcomm, Google, Huawei, Alibaba, Meta and Tesla – all have one thing in common: Arm.

Its market position was remarkable. These digital titans went to great lengths to outsmart each other, hiring the best designers and buying firms with the leading intellectual property. But jettisoning external suppliers did not extend to dropping Arm. Its architecture was too complicated and too cost-effective to reinvent and, anyway, they had invested too much in it already.

A vibrant, high-stakes market revolves around firms either licensing Arm's designs to use as basic building blocks or, by taking an architecture licence, formulating their own designs from Arm's rulebook. Its independent status appeared to have placed the company in the ultimate sweet spot. Only changing ownership once again could threaten that.

# Chapter 13

# NVIDIA'S PARALLEL UNIVERSE

## The Vision Dims

Named after the railroad tycoon Henry Huntington who opened its doors in 1914, the Langham Huntington retained the Gatsby-esque charm of America's Gilded Age. Based in an upscale neighbourhood of Pasadena, part of Los Angeles County, its palatial salmon-pink facade, marble interior, crystal chandeliers and manicured lawns looked like something from a movie set. The hotel had often been used as one, providing locations for feature films including *The Parent Trap* and *Seabiscuit* and TV shows such as *Murder, She Wrote* and *Charlie's Angels*.

Over three days in September 2019, the Langham's 23 acres swapped Hollywood glamour for the monied glitz of the tech sector. Dapper founders and executives exchanged ideas about the latest advances in artificial intelligence and quantum computing, test-drove supercars and enjoyed fine dining and performances by the singer John Legend and the mentalist Lior Suchard. And at the centre of SoftBank's inaugural Sōzō Summit – which took its name from a Japanese phrase that meant 'to imagine' and 'to create' – was Masayoshi Son, still the go-to man to bankroll any big production.

'What can you tell the founders of your portfolio companies who are doing marvellous things – what can you tell them about your learning experience?' Larry Fink, the all-powerful CEO of investment giant BlackRock, asked Son intently during a fireside chat on stage at the event. 'Because I think we actually learn more from failure than we do from success,' Fink added. Son nodded sagely and coughed. 'Obviously from your experience, my experience, the failures teach us more,' he said.[1]

The tech guru appeared to be on top of the world. Three years on from acquiring British chip designer Arm, Son had barely stopped spending. Less than six weeks after that lightning-quick deal had closed, SoftBank raised the stakes, announcing the formation of its Vision Fund to invest at least $25bn in technology over the next five years. With a potential size of $100bn, it would 'aim to be one of the world's largest of its kind', SoftBank said.

Even with Son's deep pockets, he needed to tap wealthy friends. Up to $45bn was contributed by the Public Investment Fund (PIF) of Saudi Arabia, which was eager to diversify its economy away from oil. PIF's chairman, the deputy crown prince Mohammed Bin Salman, expressed his delight at working with SoftBank and Son given their 'long history, established industry relationships and strong investment performance'.[2] Abu Dhabi's Mubadala fund, Apple and Qualcomm also chipped in.

By June 2019, the fund had amassed 81 holdings and fair value of $82bn, having backed a whirlwind of ventures, including food-delivery platform DoorDash, Indian hotels start-up Oyo and the messaging app Slack – often stumping up many billions of dollars at a time and at lofty valuations. SoftBank also transferred a 25 per cent stake in Arm into the fund, which seemingly enabled it to lessen its tax bill in Japan. It also started the clock. The Vision Fund had a 12-year horizon, 288 years fewer than Son's own vaunted vision. At some point investors would want their money back – with interest.

In July 2019, a second and larger fund, focused on the field of artificial intelligence, was announced. Based on a series of memoranda of understandings signed with backers including Apple, Microsoft, the Chinese iPhone maker Foxconn and Kazakhstan's sovereign wealth fund, expected contributions to the SoftBank Vision Fund 2 stood at an eye-watering $108bn.

---

But an embarrassing flameout was already in train. A month before Sōzō, one of SoftBank's highest-profile investments had announced its intention to float on the New York Stock Exchange. Son had backed WeWork, the temporary workspace provider led by the shaggy-haired, messianic Adam Neumann, to the tune of $11bn, although not exclusively from the Vision Fund. The initial $4bn had been conceived in typical Son fashion: based on a brief meeting and some scribbled notes.

At WeWork's last fundraising, SoftBank had pumped money into the business at a valuation of $47bn, but for the initial public offering (IPO), which was already facing postponement, it was being pitched at less than half that price. Neumann had some lofty ideas, including a company mission to 'elevate the world's consciousness'.[3] Would-be investors were rattled by a vague investment prospectus that revealed unacceptable conflicts, such as Neumann's leasing back to WeWork – now called the We Company – properties that he part-owned. Less than a week after Sōzō, having been pilloried for the revelations he smoked marijuana on board a private jet, Neumann stepped down as chief executive. The IPO was pulled soon after.

The turn of events dealt a blow to SoftBank's reputation and the thoroughness with which it carried out due diligence on its investments. In truth, its IPO programme had already delivered mixed results. Ride-hailing app Uber had listed in May that year, but on 14 August – the same day WeWork filed its intention to float – its shares slid after it posted a $5bn second-quarter loss. The performance of Slack, which floated in June, had been little better.

Not only could SoftBank not raise any money from the heavily loss-making WeWork, but it also had to put more money in so the venture could remain afloat. On 23 October, SoftBank injected up to $5bn into the business and stumped up $3bn for existing shareholders, increasing its stake to roughly 80 per cent from 30 per cent.

Two weeks later, SoftBank reported a $6.5bn quarterly loss – its first in 14 years – after writing down the value of several of its investments. 'My investment judgment was poor in many ways and I am reflecting deeply on that,' Son said.[4] Troubles with the first Vision Fund made investors jittery about the second one. Despite the raft of pledges to Vision Fund 2, only SoftBank committed. Hopes of a $108bn pot disintegrated as the high-flying company fell to earth.

Yutaka Matsuo, a professor at the University of Tokyo's graduate school of engineering who became a director of SoftBank in June 2019, reflected that his first year on the board had been 'quite a roller coaster ride' and that the Sōzō Summit 'represented the pinnacle'. Amazed by how high SoftBank's global profile was, he confessed in the group's 2020 annual report: 'I felt that perhaps it was all too good to be true.'[5]

As a survivor of the 2000 dotcom crash, Son had earned big and lost before. This time was different. On 6 February 2020, news broke that Elliott Management, the US activist investor, had taken a 3 per cent stake in SoftBank, worth more than $2.5bn, and was urging the business to reform. Identifying that

SoftBank's shares were trading at a sharp 60 per cent discount to the fair value of its assets, it encouraged Son to buy back shares and boost governance. But when Son acceded to the plan, the credit rating agency Standard & Poor's cut its outlook on SoftBank to negative, raising concerns over the impact of the buyback programme on its credit quality.

On 23 March, as the Covid-19 outbreak spread and with fears of recession looming, Son delivered his response. SoftBank would sell $41bn in assets to shore up its balance sheet that was laden with $130bn of net debt, buy back more of its shares and bonds, and help prop up some of the distressed ventures in its stable. By the time it published its 2020 annual report in July 2020, Son declared that 80 per cent of his realisations target had been met, from selling shares in his cornerstone investment in the Chinese ecommerce giant Alibaba, SoftBank's holding company and the US mobile network T-Mobile. 'I am confident in our prospects for completing the remaining 20%,' he added.[6]

Something else had already been identified to make up the numbers: Arm.

## Awkward Timing

Simon Segars also appeared at the Sōzō Summit, his smooth presentation of Arm's story so far lent extra gravitas by a newly sprouted grey beard. 'Arm is a little different relative to many of the companies here in the family,' the shaven-headed chief executive explained as he paced the stage, 'in that we are not a new company, we are not a start-up.'[7]

To illustrate the point, in the 15 minutes that Segars had to speak, Arm partners would ship 650,000 chips earning $31,700 of royalties, he disclosed. The company couldn't offer the explosive growth of some of its SoftBank stablemates, but it generated buckets of cash, some of it dating back to licences signed more than 20 years prior. With such long sightlines, when markets dived, Arm could clearly hold its value better than many more speculative assets.

Arm was intent on staying strong in its traditional markets while growing in emerging areas where it hoped that more advanced and valuable technology would enable it to increase the price it charged per chip. Its market share of mobile application processors – including smartphones, tablets and laptops – still exceeded 90 per cent in 2019. By 2028 it was aiming to have clung on to that same dominance in an advancing market it predicted would be worth $47bn. In networking equipment, Arm was aiming to take a 32 per cent market share up to greater than 65 per cent over the same time frame. Its data-centre ambition was still to be realised but furious work being done on self-driving cars meant the automotive segment showed promise too. Overall, Arm had a healthy 33 per cent share of its addressable market.

With the Sōzō crowd, Segars shared his excitement at 'being able to invest aggressively in a way that we never could have done as a public company'. Setting out those market opportunities, 'we haven't had to choose one, we are able to invest in parallel', he said.[8]

And then there was the internet of things (IOT), whose promise had captivated Son in 2016. There would be one trillion connected devices by 2035, Arm predicted. Its designs already featured in billions of microcontrollers that were installed every year in buildings, healthcare systems and traffic monitors. But the most visible sign of the company's attempt to capture the opportunity was Pelion, a software platform stitched together from two 2018 acquisitions, Treasure Data and Stream Technologies, and an existing Arm division. It was designed to simplify IOT for customers, providing basic building blocks that could manage applications, how they connected to the network and how the data that was generated was collected and processed using artificial intelligence.

IOT was still much talked about, but slow to gain traction as early hype gave way to reality. In November 2014, research firm Gartner had predicted that by 2020 there would be 25 billion connected 'things', transforming manufacturing, utilities and transport.[9] By February 2017, having predicted that three-quarters of IOT projects would take up to twice as long to implement as planned, it reined that figure in to 20.4 billion.[10,11] Problems were manifold. IOT lacked common standards, mass installation increased a corporation's vulnerability to cyber-attacks, and costs were huge.

Industry veterans weren't worried. Some ideas just took a long time to germinate. After all, the concept and form factor of the Apple Newton in 1993 didn't really translate into success until the iPad came out in 2010, 17 years later.

Arm had always gazed far ahead, trying to anticipate what end-users wanted next. But it was a year when the future looked brighter than the here and now. Born from parsimony, now the company had what felt like limitless funds and an army of engineers to shape the future. But it still had to place its bets on the right technology and focus down on delivering the advances that its customers would eventually queue up to license.

True to Son's word, Arm's workforce had ballooned by more than 50 per cent to over 6,700 staff since the takeover, including the addition of 750 people that year alone. Its vast hiring pro-gramme was a headache for rival firms in Cambridge trying to recruit at the same time. And it was 'start of season (sic) to har-vest returns' from increased research and development funding, according to its annual report that year.

Segars had grown too, emulating the Silicon Valley bosses around him. 'He's very statesman-like when he's talking externally, and he's very caring and nurturing internally,' one long-standing colleague said. He could also spout techno guff occasionally, for example describing the Arm ecosystem as 'the rallying point for tomorrow' in one company blog.

But rather than pouring more money into the firm – as was the common perception – SoftBank put in nothing beyond the 2016 purchase price, which was paid to external investors and enriched some staff. In the main, the benefit of its ownership was that Arm did not have to drain money out. No longer did it need to preserve profit margins close to 50 per cent to appease shareholders. There were some inter-company loans, but no dividend to speak of. It could run close to break-even, which let research and development spending at least double to approach an estimated £500m per year.

That was exhilarating, but in 2019 Arm still went backwards. Shipments of Arm-based chips fell for the first time in its history, to 22.8 billion, a figure later revised down further to 22.2 billion. Arm had built a broad base of applications from which it earned revenue, but it couldn't buck wider industry trends. It suffered from the weakening demand for smartphones, which still accounted for about half of its sales, as demand for 4G handsets in China slipped. A broader slowdown in chip sales brought on by the US–China trade war shrank the overall market for chips by 12 per cent to $412bn, according to Semiconductor Industry Association data. At $1.9bn, group sales that had been going up at 15 per cent before the SoftBank takeover barely rose as royalty revenues fell back.[12]

———

None of that mattered while Arm remained safe at the 'center of the center' of SoftBank, as Son had put it. But it soon became clear it might not stay there for very long at all. Soon after Sōzō, and as the WeWork debacle unfolded, Segars confirmed at Arm's TechCon event held in San Jose on 8 October 2019 that the company was looking at 2023 for an IPO, but that 'a lot of things need to fall into place' before a listing could happen.[13] Son had suggested a five-year time frame for relisting Arm at SoftBank's 2018 annual meeting, which was possible but slightly awkward for a company that worked to a 10-year product cycle. Amid flat-lining sales and heavy spending, analysts speculated the company would need to restore profits first to give SoftBank the chance to earn its money back.

In fact, an Arm IPO was considered and rejected by SoftBank in 2019 and early 2020 because the company's advisers thought it would not deliver the necessary returns. Son had been 'longing to get my hands on this hidden gem of the technology industry for over 10 years', he wrote of Arm in SoftBank's 2017 annual report. 'One day, when I look back on my long life as an entrepreneur, I believe that Arm will stand out as the most important acquisi-tion and investment I have made,' he wrote, predicting its chips would one day feature in 'running shoes, glasses, and even milk containers'.[14] However, as balance-sheet pressure mounted, the idea of an IPO quickly segued into an outright sale. It appeared that Son's ardour had waned in just three years.

Efforts to offload Arm were far less smooth than the manner of its speedy nine-week purchase. The investment bank Goldman Sachs reportedly offered it to virtually all its major customers – Apple, Samsung, Google, Qualcomm – either individually or as part of a consortium.

For the UK's tech titan, it was an unedifying spectacle. The leadership team, including Segars and his lieutenant, Rene Haas, were closely involved in touting Arm around, taking part in numerous video calls to talk up its prospects as the Covid-19 pandemic shut off face-to-face business activities.

The latest financials, industry malaise and broader economic turmoil made it a tough time to sell the business. In addition, the licensees identified as potential suitors by Goldman bankers were concerned that ownership by one of them would deter the 500-plus others. Arm had thrived by developing a unique model of serving everyone equally. Unless it could find another financial investor like itself, what SoftBank needed was someone willing to think the unthinkable. Someone with a vision to rival Son's, who could see Arm's long-term potential – and had billions to spare.

## Another Billionaire

Jen-Hsun Huang's introduction to American life did not begin with the education his parents had dreamed of back home in Taiwan. Sent on ahead of them, he and his older brother were unwittingly enrolled by his uncle into the Oneida Baptist Institute in rural eastern Kentucky, a boarding school specialising in troubled teenagers.

It was just as well his parents imbued him with determination and a thirst to learn as he was raised in Tainan, the southern Taiwan city where TSMC would eventually set up its second manufacturing base. Huang's father was a chemical engineer who had visited the US for work and was intent on moving his family there. His mother was a homemaker who, in preparation for the family's relocation, picked English words out of the dictionary for Huang and his two brothers to remember.

The challenging start did not stop Huang progressing, first as a competitive ping-pong player, and then studying electrical engineering at Oregon State University, where he met Lori, his lab partner who later became his wife.

While working at Advanced Micro Devices (AMD), he added a master's degree in engineering from Stanford University. At LSI Logic, he switched from engineering to pick up some expertise in licensing technology and hankered after running his own firm. The dream came true on 17 February 1993, his 30th birthday.

Two of his customers at Sun Microsystems, Chris Malachowsky and Curtis Priem, were put out when bosses dropped their low-end graphics accelerator, a chip designed to improve the images displayed on Sun's workstations, in favour of a high-end, unproven alternative. When a chance emerged to supply their product to Samsung, they asked the business-savvy Huang to negotiate on their behalf.

'It wasn't so much we were disenfranchised,' said Malachowsky. 'We looked at it and realised why would we do this for somebody else? Let's do it for ourselves.' They hammered out the details of their start-up in Denny's, a dingy diner in the wrong part of San Jose that 'had all the coffee we could drink', Malachowsky added.

As they worked, the trio prefixed all their files NV, meaning 'next version'. When it came to choosing a company name, they reviewed words featuring those two letters and alighted on the Latin word *invidia*, meaning 'envy'. It was something to get rivals thinking – if they ever went to the trouble of looking it up. An early press release to introduce the company was not shy in setting out their plan – nor in showing off Huang's marketing brio. 'We intend to fulfil the potential of multimedia by driving it to the limits of human perception,' he said.[15]

Compared to the flat, slow-moving images that first-generation computer gamers had to contend with in the 1970s, the next generation wanted graphics that leapt from the screen. The dash to provide chips dedicated to creating and enhancing 3D imagery rivalled the stampede in the PC market before Intel cornered the market for processors.

When Nvidia launched there were already around 30 graphics chip companies in operation. Within three years that figure rose to 70. To stand out, the trio resolved to target speed, not efficiency, in the mould of Intel rather than Arm, and not just sell on price alone. Their product was initially sold as a card that plugged into a PC to boost the speed and quality of computer graphics, reducing the CPU's workload.

Nvidia's beginnings were basic. Engineers dressed in shorts, T-shirts and flip-flops padded into a headquarters that looked like a prefab hut on a small campus in Santa Clara. The hours were long. Malachowsky turned up seven days a week but went home at six o'clock every night, returning after a family dinner. Huang attempted to work day and night, but stayed away at weekends.

The market for graphics chips was slow to take off, but two things changed. Struggling to secure production capacity with European chipmaker SGS-Thomson, Huang wrote to TSMC founder Morris Chang after he couldn't get a response from the company's US sales office. A relationship was formalised in 1998. 'They had the capacity to support our ambitions and the technology to do so,' Malachowsky said. 'And we ended up with a very close relationship.'

Then, on 31 August 1999, Nvidia claimed to usher in a new era when it heralded the GeForce 256 as the world's first graphics processing unit (GPU). In a breathless press release, Huang said its chip, which featured nearly 23 million transistors, would enable new interactive content that was 'alive, imaginative, and captivating' and would have 'a profound impact on the future of storytelling'.[16] He delighted in showing a demo of GeForce's capabilities, featuring pin-sharp marching toy soldiers, to new recruits.

By this time the company had floated its shares, which doubled in value in the first year. For Christmas 1999 it threw a big party at the waterfront Monterey Bay Aquarium, where the firm's large cohort of Indian programmers brought along parents and grandparents to share the good fortune.

Better was to come. In 2000 Nvidia was chosen to provide the core graphics chip for the first Xbox, Microsoft's attempt to muscle into the games console market that was dominated by Sony's PlayStation and Nintendo.

But there was a setback too. Excited by the Microsoft win, Huang blurted details of the contract in an all-staff email on 5 March 2000. Realising his mistake, he immediately advised everyone they were not to trade on the information. A handful did, which invited the attention of the US Securities and Exchange Commission. A two-year probe broadened to investigate Nvidia's accounting methods. It resulted in three years of restated figures – which actually boosted reported income – and the departure of the company finance director.

Another parting was the loss of Xbox as a customer in 2003. But by then Nvidia's reputation was made. In early 2001 Apple had been added to its customer list and in December 2004 a deal was signed to supply the PlayStation.

Huang's own image took shape as well. His first name was westernised to Jensen and he adopted a distinctive corporate uniform, like Steve Jobs' black turtleneck and jeans. His was all black: leather jacket, polo shirt, jeans and shoes – although the jacket got fancier over the years. Given Nvidia's performance, Huang was entitled to think of himself as a corporate rock star, but he was far from vain.

———

What supercharged Nvidia's prospects was the revelation that GPUs could be used for more than just graphics. Compared to CPUs, which were versatile and could handle a variety of tasks one after another, GPUs were workhorses that took on many similar tasks at the same time.

These relatively low-level, repetitive calculations were required to render individual pixels all at once so that images were clear and lifelike. Tailored to life beyond computer games and animations, these powers could be deployed anywhere and large numbers of commands had to be processed in parallel, chewing through great volumes of data, such as in artificial intelligence and driverless cars.

A 2009 research paper by a trio of Stanford University computer scientists, 'Large-scale Deep Unsupervised Learning using Graphics Processors', concluded that GPUs 'far surpass the computational capabilities of multicore CPUs, and have the potential to revolutionize the applicability of deep unsupervised learning methods'.[17] That meant they were ideal for machine-learning tasks that involved trying to spot patterns in reams of data and predict what came next. In some circumstances, the researchers concluded that GPUs were up to 70 times faster at cracking deep belief networks (DBNs), the giant, graphical models of data that lay at the heart of the next computing revolution.

The breakthrough brought new opportunities. Huang set his sights on supplying chips and software for data centres, the internet hubs that powered video streaming and ecommerce services remotely. GPUs were also put to work computing tasks for autonomous drones, military equipment and DNA sequencing.

There were also new rivalries. Intel sued Nvidia in 2009 over licences for technology used to make chipsets. Nvidia countersued. They settled in January 2011 when Intel agreed to pay Nvidia $1.5bn to license its technology.

Nvidia was progressing on all fronts. It tried to carry the momentum of low-energy Arm chips into PCs with a processor under the codename Project Denver. And its Tegra 2 chips for smartphones were well received by the market. Group revenues of $4bn in 2012 had transformed into almost $10bn just six years later.

As befitted its new status, the company moved into a new base a short drive from its old Santa Clara campus. Called Endeavor, the three-sided building measured 46,500 square metres (500,000 square feet) and was heavy with triangles in homage to the shape at the heart of computer graphics, including triangular skylights that made the interior bright and airy.

Before long, construction was underway next door on a larger building, called Voyager, which would feature banks of foliage on the walls and a central 'mountain' where employees could meet, work or gaze at the view.

On 8 July 2020, Wall Street observed a changing of the guard. Nvidia's stock-market value ended the day higher than that of Intel for the first time, even though its revenue was still less than a quarter of what the grandaddy of microchips reported. The ambitious Huang was playing in the big league.

---

Great corporate success bought Huang some home comforts. His house on 'Billionaires' Row' in the hills of San Francisco's Pacific Heights district had one of the US's most desirable zip codes and views across the bay to the Golden Gate Bridge. The neighbours were impressive too, including Salesforce.com co-founder Marc Benioff, Apple's former design genius Sir Jony Ive and the best-selling novelist Danielle Steel.

Huang had based himself at home during the pandemic and, sat in his swish kitchen one day during summer 2020, he received a text message from Son. 'He texts me, "Do you want to talk?" and we got on the phone, and that was about it,' Huang said, making light of the sales odyssey that Arm had been going through.[18]

Announced, oddly, on a Sunday night, 13 September, but anticipated by weeks of press leaks, Nvidia agreed to pay up to $40bn for Arm. The price marked a $9bn uplift for SoftBank after four years of ownership, but it was a modest increase given how microchip shares had rocketed. Worth about the same as Arm when it was taken private, Nvidia's valuation had soared roughly 12-fold in the intervening period. The hefty investment in Arm during the SoftBank years hadn't fed through into the price tag.

And the $40bn had strings attached. SoftBank would receive up to $17bn in cash over time, if financial targets were met, plus $21.5bn in Nvidia shares to give it perhaps an 8 per cent stake in the enlarged business. Another $1.5bn would be awarded to Arm staff.

In a statement, SoftBank said the transaction in no way changed its belief 'in the power and potential of Arm's technology' and that it created 'the world's leading computing company for the age of artificial intelligence'. Son believed that Nvidia was 'the perfect partner for Arm'. By honouring its investment commitments, the company had expanded 'into new areas with high growth potential'. The combination 'projects Arm, Cambridge and the UK to the forefront of some of the most exciting technological innovations of our time' and, as a major shareholder in Nvidia, SoftBank looked forward to 'supporting the continued success of the combined business'.[19]

The deal would combine Nvidia's AI expertise with the vast reach of Arm's processors. Because its chip designs were so pervasive, its ecosystem counted more than 13 million software programmers – an astonishing number – that Nvidia could now tap into, on top of 2 million of its own. 'Arm's business model is brilliant,' Huang wrote in a note to his staff. 'We will maintain its open-licensing model and customer neutrality.'[20]

Many pondered how it could. Some feared the very thing that had made Arm special would be lost by this latest change of ownership. Becoming part of Nvidia would certainly mark a step change. For all the talk of collaborating with other investee companies, Arm had been largely left to its own devices under SoftBank. No longer. Incidentally, the product of Son's fascination with IOT, the Pelion platform and associated assets, were separated out and would remain with SoftBank. The vision was still there, no doubt, but the business had been slow to mature.

Son and Huang knew each other well. One month after Soft-Bank's acquisition of Arm in 2016, they met in a suite of the Tokyo Conrad Hotel, located in the upper floors of the tower that housed SoftBank's headquarters. The pair were there to sign an agreement to work together on self-driving cars, but soon began thinking bigger when Son posed the question: 'The entire world will eventually become a single computer. What will you be able to do then?'[21]

Two weeks later, on 20 October 2016, Huang joined Son for dinner in California, where they sat on the terrace drinking wine and reminiscing about the Apple leader Steve Jobs, who had died five years earlier. 'At that point I had already purchased Arm and I spoke passionately about how I wanted to merge Arm and Nvidia together to revolutionise the industry by means of an AI computing platform,' Son said in one account.[22]

If that was the ultimate plan, few at Arm knew about it, and it wouldn't have been achieved by an IPO, which seemingly had been Son's preferred route for the company.

Regardless of the history, Huang was energised by the opportunity when it presented itself in his kitchen. 'I jumped on it, and by the end of the call I told him, "I will be the highest bidder. If this ever gets shopped, I will be the highest bidder." And I was.'

Huang sported a tattoo of his company's eye-like logo on his left arm, inked as a dare when Nvidia's share price hit $100. As he did the rounds after announcing the biggest semiconductor deal in history, he mused on whether it merited more body art. 'Maybe I should put Arm on my leg,' Huang said. 'I paid an arm and a leg for it.'[23]

## China Crisis

As well as talking round stunned licensees, Arm had another problem. The completion of its takeover looked as though it could be complicated by another deal done two years earlier.

On 5 June 2018, Arm announced it would cede control of its Chinese business to a group of local investors in a $775m deal. It was a move that surprised many, including former Arm executives who knew its operations in China well and thought it risky and unnecessary – and had been agreed at a headline price that fell far short of its worth to the company.

Here was a critical semiconductor market where Arm had won a strong following. Around 95 per cent of all advanced chips designed in China in 2017 were based on its technology, it estimated, and its intellectual property business there made up around 20 per cent of group revenues. Because the Chinese market was 'valuable and distinctive' from the rest of the world, Arm said it believed the transaction would expand its opportunities.[24]

The idea of a Chinese venture predated the 2016 SoftBank takeover. Arm had explored what it needed to do to maintain its popularity in China as government money flowed into the sector and state-owned enterprises took a larger role in designing silicon. So far it had fared well, despite state calls to buy Chinese where possible. But, anticipating the day when China would exceed the US as its top source of revenue, it did not want to lose ground should foreign operators fall out of favour.

After the SoftBank deal, Beijing upped the pressure on Arm. It wanted indigenous, controllable entities operating in its market, although precise guidance on how international firms should proceed was vague. Having operated in China for decades, Son was confident a spin-out could be effected that would enhance Arm's prospects. In a big win for Beijing, it sold 51 per cent of Arm China to a consortium of investment firms including Hopu,

Silk Road and Singapore's sovereign wealth fund, Temasek. The spin-out granted full access to Arm's intellectual property as it took over all its assets and employees in China and became the exclusive channel for licensing its technologies.

Arm China fell under the leadership of Allen Wu, regarded internally as a safe pair of hands and key in building up its size and value in the first place. Wu had been a member of Arm's global executive committee since January 2014 after being appointed president of Arm Greater China in January 2013. Slim and slight, he was judged to be one of the best new recruits that had joined the group in 2004 through the acquisition of Artisan.

Wu had been plucked from Arm's California sales team to become China country manager and vice president of sales for Arm China from 2009 because he was seen as more commercial than his predecessor. As a Chinese-born American citizen, he knew the market well and was an effective translator of goings-on for Arm's leaders in the UK and US. There was lingering paranoia internally that the firm might be ripped off, but he could keep track of royalties and was elevated to become president of Arm China from 2011.

Earlier in his career, Wu founded AccelerateMobile, a venture that promised to speed up the delivery of mobile data and whose team of engineers had collected experience working for Intel, Mentor, Motorola, Lucent, Oracle and Cisco. In 2001, AccelerateMobile was shortlisted in Futuredex's 'match-a-million' contest, where the winner was awarded a $1m equity investment, but it eventually faded from view.

---

Would that Arm China could have clung to a similar low profile. On 10 June 2020, Chinese media reported that Wu had been removed from his post, following the discovery, according to Arm, of 'irregularities and conflicts of interests according to whistleblowers' evidence'. But another statement, this time from Arm China, insisted that no board meeting had been convened to dismiss him and Wu continued to lead the company.

Allegations flew. Wu had secretly taken a stake in the joint venture company, one story went. He had set up a personal investment fund that invested in Arm's Chinese clients, reports said, which was already established when he brought it up at a board meeting at which its creation was not approved.

The eager lieutenant had let his new role go to his head, according to several Arm colleagues and former colleagues. He imagined he ran an independent Chinese company, Anmou Keji – translated as Arm Technology – and appointed loyalists to the top tier of management. If anyone doubted where the power had shifted, Wu brought in his own security detail too.

He was far from finished. A document seen by *Nikkei Asian Review* disclosed that Arm China planned to list its shares publicly in 2021 or 2022 and sought to surpass the revenues of its former parent company by 2025, when its sales were forecast to reach $1.89bn.[25]

As the stand-off continued, Arm China insisted its former UK parent did not have authority to remove its leader. In a post on its WeChat social media account, managers praised Wu's efforts to 'unleash and enable more industry innovations and to create true industry values'.[26]

It emerged that the high-tech business was being stymied by ancient local law. Wu had in his possession the 'chop', a wooden stamp used to validate official documents that gave him sovereign power over the company. In China's eyes, regardless of what its biggest shareholder thought, he remained the legal representative.

The row put Arm in the crosshairs of the semiconductor tensions raging between China and the West, just as it was about to come under American ownership. Huawei was its biggest customer in China.

Arm had willingly sold majority control of its most important market. Now it appeared to have lost all control.

## Pleading

The acquisition by Nvidia met with a wall of opposition. Four years after it last changed hands, history was repeating itself, except this time, because UK politicians and regulators were alive to Arm's strategic importance, there was more than just rhetoric – and the ramifications reverberated around the world.

The takeover was a topic of conversation on 27 November 2020 among thirty or so Arm veterans who convened for a virtual curry to mark the company's 30th anniversary. But the talk soon switched to catching up on what everyone had been doing since the days they fondly remembered working together.

A powerful group of tech giants including, reportedly, Google owner Alphabet, Microsoft and Qualcomm, did more than just shoot the breeze. In February 2021, they complained to the US Federal Trade Commission (FTC) about the deal. They feared that by taking control of a critical supplier Nvidia could limit other companies' access to Arm's vital designs or drive up their price. And Nvidia's ambition to expand further beyond graphics meant it would bump up against even more of Arm's 500-plus licensees when competing for business. The FTC opened an investigation.[27]

The concern was particularly acute for those working on their own Arm-based customised chips. And it was a rare moment when US worries were synced with those in China, where so many of the country's fledgling chip designers were Arm customers. If the architecture passed into US hands, they fretted that Arm could become yet another supplier blocked from selling to them in the ongoing tech trade war.

In the UK, Arm's godfather Hermann Hauser raised disquiet over the devastating impact the deal would have on the company's business model, plus jobs fears. But in an open letter to the prime minister, Boris Johnson, Hauser set out that his biggest worry was over national economic sovereignty.

Invoking President Trump's battle with China for technological dominance, Hauser warned that 'the UK will become collateral damage unless it has its own trade weapons to bargain with'. Powering the smartphones of Apple, Samsung, Sony, Huawei and more meant Arm had influence, he argued. But a sale to Nvidia would make the company subject to regulations set by the US Office of Foreign Assets Control (OFAC), which administers and enforces economic and trade sanctions based on US foreign policy and in support of its national security.

'This puts Britain in the invidious position that the decision about who Arm is allowed to sell to will be made in the White House and not in Downing Street,' Hauser reasoned. 'Sovereignty used to be mainly a geographic issue, but now economic sovereignty is equally important. Surrendering UK's most powerful trade weapon to the US is making Britain a US vassal state.'

If the deal was to proceed, Hauser sought legally binding conditions to protect the company's staff and licensees, plus an OFAC exemption. He also set out what he described as a 'natural alternative': to return Arm to the London Stock Exchange as a British-owned company, with the government as an anchor investor taking a 'golden share' to protect national economic security. If Johnson failed to do so, Hauser signed off with a warning that 'history will remember you as the person who, when the chips are down, failed to act in the national interest'.[28]

---

Over many years, the UK government had not intervened in company takeovers, no matter how large or strategically important, but now it had more grit to throw in the machine. Published in November 2020, two months after the deal was announced, the UK's National Security and Investment Bill promised to significantly expand the types of transactions where national security concerns were applicable. The roots of the legislation could be traced to the Theresa May speech, given one week before SoftBank devoured Arm, that said the UK 'should be capable of stepping in to defend a sector that is as important as pharmaceuticals is to Britain'.[29]

It was not Arm, but goings-on at another chip designer that gave the bill some impetus. The £550m purchase of struggling graphics processor specialist Imagination Technologies – incidentally, once a close partner to Arm – by Canyon Bridge was approved in a UK court on 2 November 2017, days after the Chinese investment firm's founder Benjamin Chow was convicted in the US of insider trading.

In 2020, Canyon Bridge's major shareholder China Reform attempted to install its own directors on Imagination's board but later abandoned the proposal because of political outcry. The UK had apparently waved through the deal because Canyon was regulated under US laws, but later it redomiciled to the Cayman Islands. Now fears were rising it could redomicile Imagination to China, taking its technology with it.

'The world has changed and companies – particularly tech companies – are on the frontline,' said Tom Tugendhat, chair of the House of Commons Foreign Affairs Select Committee. 'Whoever writes the code, writes the rules for the world, more than any regulation passed by bureaucrats. There's no point in taking back control from Brussels,' he added, referring to the UK's post-Brexit future, 'only to hand it over to Beijing.'[30]

Meanwhile, Arm tried to maintain business as usual. The Arm-based Fugaku supercomputer, based in the Japanese city of Kobe and created jointly by the Japanese science institute Riken and technology firm Fujitsu, was a good advert for its versatility. Ranked the world's fastest, the machine promised to advance drug discovery and climate forecasting, among other key innovations.

Trading had picked up too. Royalty revenues bounced 17 per cent in 2020 thanks to strong demand for 5G smartphones. And Arm still had an eye on the future. On 30 March 2021 it unveiled ARMv9, the ninth version of its architecture and the first upgrade in a decade. Segars confidently predicted it would be at 'the forefront of the next 300 billion Arm-based chips'.[31]

Nvidia was busy displaying its capabilities too. That summer, it announced that Cambridge-1, the product of a $100m investment by the company, would become the UK's most powerful supercomputer, regardless of whether its takeover went through. Located in Arm's backyard, as the name suggested, the machine would enable top scientists and healthcare experts to use the powerful combination of AI and simulation to accelerate digital biology work and boost the life sciences sector, such as better understanding brain diseases. Projects were already lined up with AstraZeneca, GSK and Guy's and St Thomas' NHS Foundation Trust.

---

On 19 April 2021, the Secretary of State for Digital, Culture, Media and Sport, Oliver Dowden, issued a public interest intervention notice (PIIN), confirming that he was intervening in the sale on national security grounds.

Soon after, the Competition and Markets Authority (CMA) said the deal raised serious competition concerns by restricting access to Arm's intellectual property that could push up prices of products ranging from cars to mobile phones and data centres. Andrea Coscelli, the CMA's chief executive, said it could create problems by 'ultimately stifling innovation across a number of important and growing markets' and an in-depth investigation followed. Chinese regulators were combing over the deal too.

None of this was good news but perhaps the final straw came when the FTC sued to stop the transaction on 2 December. The FTC's bureau of competition director, Holly Vedova, explained it was acting 'to prevent a chip conglomerate from stifling the innovation pipeline for next-generation technologies' amid concerns the deal 'would distort Arm's incentives in chip markets and allow the combined firm to unfairly undermine Nvidia's rivals', particularly in the areas of autonomous cars, data-centre servers and cloud computing.

This lawsuit 'should send a strong signal that we will act aggressively to protect our critical infrastructure markets from illegal vertical mergers that have far-reaching and damaging effects on future innovations', Vedova added.[32]

Ironically, the further stock-market investors thought the deal was slipping from Son's grasp, the more valuable it potentially became to him. Nvidia's soaring share price inflated the paper element of their agreement, valuing Arm at $80bn at its peak. But that was little comment on Arm's prospects, merely rising hopes that a distraction was about to be removed for Nvidia's leadership. Whatever the price tag, the UK's technology champion was proving too valuable to be sold.

---

Arm and Nvidia pleaded. The last roll of the dice came in their joint submission to the CMA on 20 December 2021, when the picture painted of Arm was not that of a thriving tech business with hundreds of partners that shipped billions of chips, but of a severely challenged company in desperate need of support. SoftBank's investment had intended to spur growth in the data-centre and PC market to better compete with Intel and AMD but its success in these areas was far from proven.

This was clearly not the right forum for celebrating Arm's strengths. Rather, it was the moment to highlight its weaknesses. Arm's limitations as a standalone entity that solely licensed intellectual property 'have become apparent over many years', the submission said.

Its original mobile market, still the largest source of revenue, 'is saturated'. The data-centre and PC markets 'are far more difficult to crack'. Arm was at a disadvantage to Intel and AMD, who made profits from several levels of the technology stack. Lacking expertise, scale and resources, it 'could not generate the revenue necessary to invest and compete toe-to-toe with the entrenched x86 incumbents'.[33]

The question of why Arm did not push up its royalty rates was not a new one. Quizzed on the subject around the time he became chief executive in 2013, Segars responded: 'We could do that and we could probably enjoy some more revenue for some time, but our customers would go off and do something else or have less healthy businesses. If we tried to extract lots of money out of the ecosystem, we'd have less companies supporting the Arm architecture and that would limit where it could go.'[34]

Alternatively, the submission said a sale to Nvidia would 'materially change Arm's incentives and opportunities', which was surely a reference to the $1.5bn of Nvidia equity due to be dished out to staff on completion. That worked out at $230,000 for each member of its 6,500-strong workforce, but of course the lavish rewards would be skewed towards the firm's leaders and star engineers.

Those that cherished Arm's independence should be mindful that 'soaring profits for Intel, Apple, Qualcomm, and Amazon have not manifested as a "win" for Arm, or for competition in general'. The CMA document painted the failure of the Nvidia deal as a doomsday scenario, leaving Arm at the mercy of 'unsentimental' public markets that would demand 'profitability and performance' and no doubt strategic changes and cost-cutting. An Arm IPO, considered and rejected over the last two years, would 'narrow its focus and limit investments'. In particular, Nvidia was concerned that Arm would be forced to focus on mobile and IOT at the expense of designs for PCs and data centres.

Nor did Arm portray itself as a share worth buying. When set against the recent 'skyrocketing revenue growth and profits, as well as soaring market valuations' of Apple, Qualcomm and Amazon, Arm had suffered from 'comparably flat revenues, rising costs, and lower profits that would likely present challenges for a 30-year-old public company'. In the document, Segars explained: 'We contemplated an IPO but determined that the pressure to deliver short-term revenue growth and profitability would suffocate our ability to invest, expand, move fast and innovate.'[35]

The pleas were not enough. On 8 February 2022, both sides bent to the inevitable as the deal was abandoned because of 'significant regulatory challenges' and despite 'good faith efforts by the parties'.[36]

Huang declared, 'Arm has a bright future, and we'll continue to support them as a proud licensee for decades to come.' Son appeared to have no regrets, declaring that Arm 'has entered its second growth phase'. SoftBank would prepare Arm for IPO no later than March 2023 to 'make even further progress', he added. Segars' declaration made at Arm's TechCon in October 2019 looked like coming true after all. SoftBank had simply gone the long way round.

But it wouldn't be Segars leading the company back on to the public markets after seven years away. He stepped down immediately, to be replaced by his close ally, Rene Haas, who had served as president of Arm's IP products group (IPG), its main division, since 2017.

'Arm has defined my working life,' Segars said in the statement, 'and I am very thankful for being given the opportunity to grow from graduate engineer to CEO.' He added: 'I'm very bullish on Arm's future success under Rene's leadership and can't think of anyone better to lead the company through its next chapter.'[37] The disappointment was palpable.

## Chapter 14

# THE RISK FROM RISC-V

## The Switzerland of Semiconductors

The 'Schiffbau', meaning shipyard, once welcomed hundreds of engineers every day, but now the striking brick building with tall, arched windows served as a theatre and dining venue. It was one of the few remnants of west Zurich's industrial past not to have been swept away as its surroundings were transformed into a fashionable district to live and work. Another, on the nearby Schiffbaustrasse, was a tower that was part of a long-forgotten blast furnace whose noisy clank had been heard from afar.

Between the two landmarks in Industriequartier in Switzerland's largest city stood an anonymous grey office block, recently refitted to offer modern, contemporary desk spaces for the influx of white-collar workers, complete with a cafeteria on the attic floor. In late 2019, another organisation gave notice that it would be joining, in a manner of speaking, tenants including the Swiss law firm MLL Meyerlustenberger Lachenal Froriep and BDO, the accountants, in Schiffbaustrasse 2.

RISC-V International was a world away from the engineering heritage of the area – although its partners were involved in manufacturing that was as advanced as it got. But this venture made nothing, generated little noise from its operations and employed next to no one locally. It had a registered address, which was a place to pick up the mail should any ever be sent, and that was about the extent of its presence.

The chip designer Arm was often described as the 'Switzerland of semiconductors', offering design licences to all that would pay for them while maintaining a studied neutrality among sworn enemies.

RISC-V was the new kid on the block, seeking to do the same but going further. Not only did it officially locate itself in Switzerland as it took shelter from the ill wind of geopolitics blowing through the microchip industry, but it intended to offer everything that Arm did, only for free.

And it was working. In four years since its inception the foundation had attracted around 325 fee-paying members, including most of the top chip suppliers: Qualcomm, automotive specialist NXP Semiconductors and, from China, Alibaba and Huawei. Just as TSMC, the foundry pioneer, had succeeded in lowering barriers to entry by removing the need to build a foundry, RISC-V was a magnet for chip-design start-ups on tight budgets. It sparked a new round of innovation and excited many in China and Russia, who were eager to try an architecture other than the West's Arm and Intel, home of x86.

RISC-V also found favour in the European Union, where the European Processor Initiative (EPI) pinned hopes on it as one path to semiconductor independence for the bloc. By 2023 the EPI aimed to build a new supercomputer based on RISC-V designs.

What turbocharged industry interest further were the stories that emerged through 2020 about Nvidia considering the acquisition of Arm. If Arm's independence was about to be compromised – even though Nvidia CEO Jensen Huang said it would not be when a deal was eventually announced – there was no better time to explore the open-source, royalty-free back-up.

Analysts speculated that the proposed $40bn takeover would actually erode Arm's value, not crystallise it, as it helped to establish RISC-V as a credible alternative that could be used by all chipmakers but owned by none.

In the pair's December 2021 submission to the Competition and Markets Authority (CMA), Arm and Nvidia paid RISC-V the ultimate compliment: 'RISC-V has two potential advantages over Arm today—it is both less expensive and more customizable than Arm,' they wrote. 'Even if customers prefer Arm today, RISC-V creates a very real competitive constraint.'[1]

In fact, one customer that knew RISC-V intimately was Nvidia. It had been a founding member when RISC-V International was formed in 2015 to oversee the new standard, and soon after adopted its architecture for a new graphics processing unit (GPU) microcontroller, with the expectation of vastly improving performance.

Clearly the biggest chipmakers liked to hedge their bets when new technologies emerged. But Arm could not be complacent. In past battles, giant corporations had eyed its lucrative franchise, but here was a free competitor that was available to everyone, everywhere.

## Back to Berkeley

RISC-V might never have existed if Arm had been willing to let Krste Asanović experiment with its instruction set. The professor in the electrical engineering and computer sciences department at the University of California, Berkeley, clearly recalls the conversation sat in a café in San Francisco with an Arm representative.

For a new project, he wanted to play around with Arm's designs and share his ideas with fellow academics. Asanović was particularly put out that he couldn't even build a simulation of some instruction-set specifications available to download from Arm's website.

'There's a history of Arm being too protective, which makes sense given the business model,' Asanović said. 'Intellectual property is their crown jewel so they are not going to give it away. But on the other hand, that helped drive RISC-V.'

Asanović was raised in Corby, an unremarkable Northamptonshire town, after his parents moved to the UK as refugees from the former Yugoslavia following the Second World War. As part of the first generation for whom computer programming became a hobby, he persuaded his parents to buy him an Acorn Atom. And when Arm history was being made, he was in the vicinity.

During the years Asanović was an undergraduate at Cambridge, studying electrical and information sciences, elsewhere in the city Steve Furber and Roger Wilson popped champagne corks with expectant Acorn executives after demonstrating the first Arm version. Of course, they only resolved to push on with their own designs after Intel declined to let the firm license and modify its 16-bit 80286 chip for what would become the Acorn Archimedes.

However, Asanović's connection to Arm's heritage grew closer in the years after he left the UK. Soon after he got to Berkeley in 1989, he took the computer architecture class led by David Patterson, whose works on a reduced instruction set computer (RISC) had been delivered to Wilson and Furber by Andy Hopper, the Cambridge lecturer and Acorn director. Patterson's findings inspired the flurry of RISC architectures that emerged during the 1980s, from which Arm became the clear market leader a decade or so later. Asanović began working closely with him.

Patterson, a successful powerlifter despite his advancing years, became the director of Berkeley's Parallel Computing Laboratory (Par Lab), a research centre founded in 2008 from which RISC-V grew. It had been set up with $10m from Intel and Microsoft after being judged as the top proposal in a competition among 25 leading computer science university departments. Par Lab later attracted further third-party funds, including from the California state.

A new instruction set wasn't why the contest had been launched. Long-standing 'Wintel' partners Intel and Microsoft had grown concerned in the aftermath of Intel's Pentium 4 failure. Launched in 2000, the processor had a higher clock speed than had gone before but it was expensive and hot and didn't perform as well as competing models. As processors with several cores were introduced, the mission had been to develop 'parallel' software that enabled complex tasks to be broken down and worked on simultaneously.

Asanović's team had previously used MIPS, the Sun Microsystems-originated SPARC or x86 instruction sets in its research projects, but Arm had become extremely popular in the meantime. The problem they kept encountering was that when they made changes to an architecture, the software they ran no longer worked, so they considered going back to basics by coming up with an instruction set of their own. Coupled with that, Arm wouldn't let them experiment or, if they had paid for a licence, share findings with other, less well-resourced universities that couldn't afford a licence of their own. Finally, the team preferred to work in 64-bit whereas Arm had not yet progressed on from 32-bit.

'One of the greatest powers in the universe is grad-student naivety,' said Asanović, paying tribute to colleagues Yunsup Lee and Andrew Waterman. 'When you don't realise something is impossible you try it anyway.'[2]

When Arm's 64-bit v8 instruction set eventually came out in 2011, Patterson doubted the need for an alternative. But when he took a close look at it, after his publisher asked him for an updated version of his widely used textbook, *Computer Organization and Design*, he thought it more complex than first imagined. So hard was it to write a textbook based on the 5,000-page Arm manual, Patterson chose to write on just part of it, which he dubbed 'LEG', short for 'leave out extraneous garbage'.

The RISC-V framework 'wasn't the research goal, it was just a thing we built ourselves – the scaffolding we needed to do our work', Asanović said. It began with just 40 unique instructions – devastatingly simple by necessity because it was built by only a handful of people. Using the RISC name for what they counted as the fifth iteration of the technology was about reclaiming something that had only ever been intended as the title of the original Berkeley project. 'Dave regretted it becoming a generic name like Hoover,' Asanović added.

Berkeley's ASPIRE Lab succeeded Par Lab and led to several RISC-V compatible microprocessors being built. Support came from the Pentagon's Defense Advanced Research Projects Agency (Darpa). By now Asanović and Patterson sensed there was wider interest in what they were doing. What had initially been designed for research and education could have commercial uses too. By 2014, with their base design complete, they decided to test the market by publicising RISC-V.

Even though cut-throat competition was commonplace, great swathes of the computing and communications industries were based on open standards available to all. There was the computer wiring technology Ethernet used in local area networks, Bluetooth, a short-link radio technology developed for wireless headsets and Linux, a computer operating system released in 1991 as an alternative to Microsoft Windows, whose uses had spread to supercomputers, cars and, as the basis for Google's Android mobile operating system, most smartphones.

———————

In an August 2014 research paper entitled 'Instruction Sets Should Be Free: The Case For RISC-V', Asanović and Patterson wrote that there was 'no good technical reason' for the lack of a free, open instruction set architecture (ISA) for hardware. They criticised the often drawn-out licence negotiations and costs that restricted academics and others who worked with tiny volumes from using Arm and others. The paper concluded that 'the case is even clearer for an open ISA than for an open OS (operating system), as ISAs change very slowly, whereas algorithmic innovations and new application demands force continual OS evolution'.[3]

The same month they took a table at Hot Chips, an annual conference on high-performance chips that took place over three days at the De Anza College in Cupertino, California. Asanović thought there was some frustration with Arm, centred around the amount it charged in royalties. The company's own pleading of 'comparably flat revenues, rising costs, and lower profits'[4] when set against the riches of Apple, Qualcomm and Amazon did not wash with start-ups for which spending several million dollars on licensing was barely in reach.

But cost wasn't all. Asanović met start-ups who recounted stories of the two years they had been trying to negotiate a licence with Arm and the lack of flexibility over what the company would permit them to do when they secured one.

Arm had succeeded better than any other company in commercialising RISC, but the plucky, cash-strapped start-up had long ago transformed into an establishment provider that developers delighted in moaning about. RISC was not Arm's innovation

– that was to license chunks of intellectual property that saved chipmakers from having to design processors for themselves.

RISC-V's innovation was to make that segment of the industry free. It had not been created to take over the world, but now its founders wanted to see if the world would take it over. Just as Linux had been a kernel for open software, RISC-V could be a kernel for open hardware: a set of instructions used by all chipmakers and owned by none of them so that it could be freely modified.

At a workshop it held several months after Hot Chips, 40 companies attended, interested in making use of the architecture. They had one concern: RISC-V's home was within a university, where academics could get distracted by competing projects and graduates moved on. It wasn't a stable enough environment for them to invest millions of development dollars in. Asanović and Patterson resolved to spin out RISC-V into a foundation that generated revenue purely through membership fees. The Linux Foundation, an obvious model, took care of its back-office functions.

Its birth tapped into some frustration that Arm hadn't done more with its RISC leadership. In a 2015 interview, David Ditzel, one of Patterson's former RISC collaborators, damned with faint praise Arm's 64-bit architecture. 'It's a very generic RISC,' he said, 'and I think it's a really excellent RISC architecture, but it could've been done in the 1980s.'[5]

## Retaliation – and Relocation

Like an elephant trying to swat a fly, on 9 July 2018 Arm resolved to do something about its new competitor. The company launched a website, riscv-basics.com, designed to plant seeds of doubt in the minds of developers who might use RISC-V as their processor architecture instead of Arm. Under an Arm logo and the heading 'RISC-V Architecture: Understand the Facts' it set out several factors to consider carefully.

Would-be customers should not be swayed just because RISC-V had no licence fee or ongoing royalties, it suggested, because that 'accounts for a small fraction of the total design-to-delivery investment required to create a commercial processor'. A large ecosystem of partners 'guarantees market choice, product quality and an optimal time to market' but RISC-V ecosystems 'have not yet reached this stage of development', it added. As for security threats, 'RISC-V based products are relatively new and have yet to benefit from years of scrutiny from partners and industry experts'.

In conclusion: 'Whether you are looking to create a chip from scratch or looking for a complete solution, take advantage of an architecture that has been tried and tested in more than 125 billion chips and already in processor designs licensed by more than 500 partners.'[6]

This campaign, characterised as one of fear, uncertainty and doubt (FUD) in the industry, was not uncommon among US software firms but these aggressive tactics had never been Arm's approach. Now it was the one facing the threat of a nimble upstart, someone in the corporate machine must have thought it was a good idea. The website was taken down a day later, after uproar from angry Arm engineers in Cambridge.

———

The first international concerns Asanović picked up over RISC-V's domicile came from India. It was several years before President Trump's microchip war with China, but some developers had long memories. In 1998, the US imposed sanctions on India after it detonated three underground nuclear explosions. Efforts to deter the spread of nuclear weapons resulted in restrictions being slapped on computer exports and solidified India's focus on building its own capabilities.

As the geopolitical heat increased so did the intensity of conversations. Unlike in the years when Linux rose to ubiquity, RISC-V was trying to open itself to all at a time when nations' mistrust of each other was rife. RISC-V was contained within a US foundation, and an open standard was not subject to export controls. Still, there was confusion.

In June 2019, more than two dozen standards groups – including those that oversaw SD memory cards and Ethernet – wrote a letter to the US Commerce Secretary Wilbur Ross asking for clarification of the rules on working with Huawei, the Chinese telecoms group that Trump had turned into a corporate pariah. They warned that the Huawei restrictions posed a risk that could send standards-setting overseas.

As RISC-V chairman, Asanović listened to worries from partners in Japan, China and the EU. To underscore its neutrality, the foundation announced in November 2019 it was moving to Switzerland from Delaware so that it wouldn't fall foul of White House trade curbs.

Calista Redmond, the foundation's chief executive and an IBM veteran, said its global collaboration had faced no restrictions to date but members were 'concerned about possible geopolitical disruption'.[7] The move was not based on 'any one country, company, government, or event' but 'reflective of community concern and managing strategic risk for our community investing in RISC-V for the next 50+ years', the organisation said.[8]

In search of neutrality, Switzerland was obvious, but the decision was made easier because there was a very active group of RISC-V developers at ETH Zurich, a public research university.

But the move went down badly in the US, where its leadership physically remained at offices in San Francisco. Arkansas Republican Senator Tom Cotton said shifting the foundation to ensure it could retain Chinese members was 'short-sighted at best' and added that 'if American public funds were used to develop the technology, it's also completely outrageous'.[9]

## The Funds Flood In

The creators of RISC-V were also in the vanguard of commercialising it. SiFive, the first fabless semiconductor company to build customised silicon based on the RISC-V architecture, was co-founded by Asanović, Lee and Waterman.

The venture attracted speculative dollars from some big names. In August 2020 SiFive raised another $61m in funding from investors including SK Hynix, South Korea's giant maker of memory chips, and Prosperity7 Ventures, an investment subsidiary of Saudi Arabia's state-energy major Aramco. That made $185m of funds raised overall since 2015, including from Intel, Qualcomm and Western Digital.

In September 2020, coincidentally a few days after Nvidia announced it was acquiring Arm, the company underlined its intent by hiring former Qualcomm executive Patrick Little as its chief executive.

'The industry is transforming away from general-purpose computing to something domain-focused,' Little said. 'With the news this week, it's now accelerating, and the magnitude has really picked up. Now there are many companies saying it's time to look at open versus closed solutions.'[10]

And in March 2021, on the day the mighty Intel announced it was changing direction to manufacture other designers' chips by adopting the 'foundry' model popularised by TSMC, SiFive said it would work with it to make RISC-V designs more widely available.

One year later, soon after Arm's sale to Nvidia collapsed, SiFive raised another $175m to be valued at $2.5bn, and began talking about the prospect of an IPO. Its P550 chip design was aimed at challenging Arm's dominance in the smartphone arena. And, as part of its mission to build chips no matter what architecture customers chose, Intel set up a $1bn innovation fund with some focus on boosting RISC-V.

A technology that began by powering cheap, low-risk functions showed signs of trading up. In September 2022, SiFive was selected by NASA to provide the brains for chips that would power future space missions.

The company had funds and customers, but it needed staff. It drove its tanks further onto Arm's lawn with the opening of a new research and development centre in Cambridge, alongside a commitment to hire more than 100 new employees as it staffed up internationally. To emphasise it was putting down roots – and its eagerness to hire locals – soon after it struck a sponsorship deal with Cambridge United Football Club, a third-tier team whose 'success on the pitch, with fewer resources than their competitors, mirrors the SiFive success story', SiFive said.[11] Asanović, the local university alumnus, paid a visit to the club to explain how RISC-V worked.

Overall, it was difficult to know precisely how prevalent RISC-V chips had become, because the foundation asked firms to disclose usage voluntarily and many chose not to. Redmond estimated in summer 2022 that at least 10 billion cores had been shipped. Its membership had ballooned to become more than 3,100-strong across 70 countries. Arm's partners – many of whom also utilised RISC-V – shipped the equivalent figure every four months. The Arm alternative still did not have the performance record, the vast software ecosystem or the years of application development behind it. But it was certainly a meaningful start.

Semico Research predicted that by 2027 there would be 25 billion RISC-V-based artificial intelligence enabled systems-on-a-chip (SoCs) on the market that combined all the necessary electronic circuits for a single device. Another market watcher, Counterpoint Research, thought RISC-V could achieve a 25 per cent share of the internet of things (IOT) segment by 2025, where simple, low-powered sensors were required.[12]

'I would put relations at neutral,' Redmond said when asked about how RISC-V International and Arm, which had dabbled by open-sourcing designs to a few of its processor cores, got on. 'We don't have any active engagement or projects between us but there isn't discord between us either. We think that the pie is big enough for all.'

Asanović was more pointed, saying in RISC-V's tenth anniversary video, 'I will be curious to see which other ISAs are left' a decade down the line, 'or if something new comes up'.[13] Battle lines were drawn.

Arm had made headway in data centres and proved its worth to every giant tech firm, but the nature of innovation meant it could not stop looking over its shoulder.

## Chapter 15

# GOING GLOBAL: SOME FURIOUS SOVEREIGN SPENDING

### The Broken Nest

In the grand hall of the Palais de Chine Hotel in Taipei's Datong district, red tortoise cakes were piled high. Stacked into two towers, the traditional oval-shaped pastry, consisting of a sticky rice-flour skin encasing a sweet filling, symbolised prosperity and longevity. On 29 November 2021 the cakes formed the centre-piece for a very special private celebration.

The hotel, adorned with European antiques and velvet swags, welcomed the Taiwanese President, Tsai Ing-wen, plus business leaders including the chair of the wireless and TV chip supplier MediaTek, Tsai Ming-kai, and Terry Gou, the founder of Foxconn Technology, best known for its manufacture of the iPhone. Dozens more dignitaries joined them beneath the hall's opulent copper-foil ceiling that was studded with 16 crystal chandeliers.[1]

They had gathered to mark a significant milestone: the 90th birthday of Morris Chang. The white-haired founder of Taiwan Semiconductor Manufacturing Company (TSMC), Taiwan's largest and most important company, was actually born in July, but the Covid-19 outbreak prevented celebrations until this moment.

Chang had retired from TSMC in 2018 but his impact on the company and the nation had not diminished. And even without him at the helm, TSMC could celebrate its own prosperity and longevity with some confidence.

Over the years, everything had got bigger – except the process technology with which the company produced chips. Building a new fab typically cost $20bn – 10 times the bill from a generation ago. But TSMC had the means to afford the relentless investment. In 2021, revenues soared 19 per cent to $57bn and net income was up 15 per cent to $21bn, as demand for 5G and high-performance computing chips increased. Stripping out the simpler memory-chip production that it had never focused on, TSMC commanded a 26 per cent market share.[2]

It was still pushing boundaries. Remarkably, half of its income came from chips made at the 7nm process node and below, which first went into commercial production only three years earlier. That was on track to rise to 70 per cent by 2023. At its Fab 18 facility in Tainan, southern Taiwan, volume production at the 3nm node began in late 2022.

All this as the country still battled incursions from across the Taiwan Strait. In March 2021, Taiwan prosecutors charged Bitmain, a Chinese crypto and AI company, with setting up two front companies to poach staff. Chinese firms were permitted to set up branches in Taiwan but they had to be registered with the government and their activities were closely monitored – especially if efforts to export microchip technology were suspected.

Soon after, Taiwan's labour ministry banned recruitment companies from hiring for jobs in China. The measure was because 'China has become more aggressive in poaching and targeting top Taiwanese chip talent',[3] a notice from the ministry said, with stiffer penalties if the recruitment involved semiconductors and integrated circuits.

In a rare public outing, Chang cast doubt on how effective China's sustained effort to capture industry leadership had been. On 21 April 2021, at a forum hosted by the *Economic Daily* in Taipei, he said that 'mainland China has given out subsidies to the tune of tens of billions of US dollars over the past 20 years but it is still five years behind TSMC'. In the design of logic chips, China's capability was still one to two years behind the US and Taiwan, he added. Chang's message was clear: 'The mainland is still not yet a competitor.'[4] Its biggest rival remained Samsung, he claimed. And yet they were closely bound together. TSMC relied on China for 10 per cent of group revenues in 2021.

There was no let-up. On 9 March 2022, some 100 Taiwan officials raided Chinese chip and component suppliers at 14 locations across four cities over allegations they had been illegally poaching Taiwanese workers. The investigation reportedly hauled in 60 Chinese nationals suspected of lifting trade secrets or luring talent. Chinese headhunters were specifically looking for experts in integrated circuit design, electronic design automation, telecommunications, and electric vehicle manufacturing, the local news outlet UDN said.[5] Further raids took place in May.

---

Taiwan and TSMC stood strong despite a backdrop of increasing political tensions. Given President Trump's combative approach to international affairs, it was hard to imagine that US relations with China could worsen after Trump left office. But they had.

In January 2021, during the first few days of Joe Biden's presidency, Taiwan reported a 'large incursion' of Chinese warplanes into its airspace, followed by more in April, interpreted as a test of US commitment to the region. US Admiral John Aquilino, head of the Pentagon's Indo-Pacific command, issued the sombre warning that a Chinese invasion of Taiwan was 'much closer to us than most think'. China's President, Xi Jinping, said that 'reunification' with Taiwan 'must be fulfilled'.

Taiwan had watched in horror in June 2020 as China introduced a national security law in Hong Kong that had the effect of reducing autonomy. The former British colony was supposedly run under the model of 'one country, two systems' that had once been offered by the mainland to Taiwan too. A few months earlier, the Taiwanese President, Tsai Ing-wen, won re-election after polling a record 8.2 million votes, a result that was a firm rebuke to China.

Biden strengthened export restrictions that were designed to stymie China's semiconductor progress. Shipments by chip-making tool suppliers KLA Corp, Lam Research Corp and Applied Materials were curbed, while Nvidia and AMD were ordered to stop selling their top artificial-intelligence microchips to China for fear they may find a 'military end use'. Another firm that looked to be in the line of fire was China's leading maker of memory chips, Yangtze Memory Technologies Co (YMTC).[6]

A research paper published in *Parameters*, the US Army War College quarterly journal, suggested a radical response to defuse the situation and emphasised the strategic importance of Taiwan's microchip industry. The 2021 report, led by Jared McKinney, chair of the Department of Strategy and Security Studies at the eSchool of Graduate Professional Military Education, Air University, proposed that Washington and Taipei should pursue a 'broken nest' strategy of deterrence by convincing China they would destroy TSMC's facilities in the event of an invasion.

'An automatic mechanism might be designed, which would be triggered once an invasion was confirmed,' McKinney and his co-author, Peter Harris, wrote. The US could also announce contingency plans with Taipei 'for the rapid evacuation and processing of the human capital that operates the physical semiconductor foundries'.[7]

If they remained in the country, industry executives doubted TSMC's Taiwanese engineers would work under Chinese rule. And even if the plants were not bombed – by either side – there would be untold disruption to make the post-pandemic chip shortages look inconsequential. But then, with a world war raging, there would be more at stake than keeping up with demand for the latest car model or games console.

Political interventions in the industry stepped up. In April 2021 the Italian prime minister Mario Draghi halted the sale of Milan-based LPE, which produced components for power electronics applications, to a Shenzhen company. The following February, the $4.9bn takeover of German chip supplier Siltronic by Taiwan's GlobalWafers collapsed when it did not receive regulatory approval in time in Germany. Meanwhile, the Russian invasion of Ukraine threatened new havoc for the semiconductor supply chain because half of the world's neon, a gas used in lithography, was produced in Ukraine as a by-product of steelmaking.

In May 2022, on his first presidential tour of Asia, Biden insisted that his country's policy towards Taiwan had not changed but warned that China was 'flirting with danger' over the island. He vowed to intervene militarily if it was attacked. In the same month, a 6.1-magnitude earthquake struck off Taiwan's east coast. It was as if nature was providing a soundtrack to the dire situation.

## China Keeps Trying

After almost three years of diplomatic wrangling over her fate, Huawei's chief financial officer, Meng Wanzhou, was allowed to leave Canada on 25 September 2021. Standing outside British Columbia's Supreme Court, wearing a black dress with white polka dots, she removed her facemask to read from prepared notes to reporters in halting English.

'Over the past three years my life has been turned upside down,' Meng said. 'It was a disruptive time for me as a mother, a wife and a company executive but I believe every cloud has a silver lining. It really was an invaluable experience in my life.'[8]

To break the impasse, a deal had been struck in which Meng did not plead guilty to fraud but did admit some information about the company was misrepresented in an effort to avoid certain US sanctions. It was enough for the US Department of Justice to drop its extradition request. At the same time, the two Canadians, Michael Kovrig, who was awaiting sentencing, and Michael Spavor, who had been found guilty of spying and sentenced to 11 years in prison, were released.

Meng touched down at Shenzhen's Bao'an International Airport to a hero's welcome. It was a happy homecoming for the daughter of Huawei's founder, Ren Zhengfei, but far from the end of the story for the firm that still faced US charges of racketeering and conspiracy to steal trade secrets. The cost of operating in the crosshairs of the increasingly vicious battle between the US and China for microchip supremacy was beginning to tell.

There had been surprising successes, however. In mid-2020, Huawei had become the world's biggest seller of smartphones – the first quarter in nine years that a company other than Samsung or Apple had led the market. There were extenuating circumstances, namely that China came out of the Covid-19 lockdown ahead of the rest of the world. While the previous industry leader Samsung was hit by plunging sales in many major markets, Huawei consolidated its home advantage to sell more than 70 per cent of its smartphones in mainland China, according to estimates.[9]

But, still languishing on the US entity list, the general trend was down. In 2021, Huawei's revenues fell 29 per cent to 637bn yuan ($100bn), the first full-year drop it had recorded, as the export clampdown bit.[10] Huawei's rotating chairman Guo Ping said in a New Year letter to employees that 2022 'will come with its fair share of challenges' but that he was satisfied with business performance to date. 'An unpredictable business environment, the politicisation of technology, and a growing deglobalisation movement all present serious challenges,' Guo wrote.[11]

The company appeared to be gradually disintegrating. After Google was forced to withdraw its licence for the Android operating system, Huawei sold its youth mobile-phone brand Honor to a Chinese consortium of agents and dealers, including the government of Shenzhen, with the hope that the Google relationship could be restarted. Unable to buy x86 chips from Intel, its computer server business was hived off too.

However, reports of Huawei's death were premature. During the Trump administration, licences worth $87bn for Huawei were approved after its blacklisting, although $119bn were denied, according to US Commerce Department documentation.[12] For example, as the company's stockpile of Kirin chipsets ran out, Qualcomm obtained permission to sell to Huawei its older 4G chips that were deemed to be no threat.

In a sign of its resilience, at the same time as sales crashed, Huawei profits rose 76 per cent to 114bn yuan ($17.8bn). The company raced to replace revenues lost in its consumer and 5G equipment businesses by driving deeper into areas such as energy – where its kit converted the electrical output from solar panels – mining and cars. They were sectors that did not require the most advanced chips, so by making do with the 28nm process node and higher Huawei could compete without recourse to overseas production, or else it found European suppliers with which to trade.

The company made vigorous attempts to build what it was no longer allowed to buy. Shorn of Android, it developed its own mobile operating system, HarmonyOS, to run on the remaining handsets it sold.

Through a venture capital arm, Hubble Technology Investment, in 2021 it invested or increased its holding in 45 domestic companies, more than double the 2020 number, which operated at various stages along the microchip supply chain, including packaging.[13] Bloomberg reported in July 2022 that HiSilicon, its chip design arm, was hiring dozens of engineers to develop its own semiconductor design software, a niche field dominated by two US providers, Cadence Design Systems and Synopsis.[14]

Blocked from using TSMC for manufacturing, Huawei's biggest challenge was to create its own processor production line. It was not something it shied away from. Reports suggested it was working with Fujian Jinhua Integrated Circuit Company to revive its factory in the southern Chinese port city of Quanzhou. The plant had been inactive for several years ever since the memory-chip maker was accused of stealing US trade secrets. To compound the external cool Huawei was trying to project, in December 2022 the firm's latest chairman, Eric Xu, declared US restrictions were 'our new normal, and we're back to business as usual', as he reported flat annual sales.

---

The story was similar across the entire Chinese semiconductor industry. Rather than beating a nation into retreat, US restrictions spurred renewed efforts to become self-sufficient.

There was heavy irony here. Local customers had preferred to import chips, until they had little choice but to buy locally. The US was in a race to isolate China from the advanced chipmaking industry before it could replicate it. But its actions were stimulating domestic demand and the increasing number of companies frozen out at all points of the supply chain were learning how to make do from the crackdown's poster child, Huawei.

The Mercator Institute for China Studies (Merics), a Berlin-based research unit, noted in late 2021 that export controls might have damaged China's leading chip design house HiSilicon, 'but have not had an observable negative impact on China's larger chip design ecosystem'. In fact, it suggested, 'they might boost this ecosystem by dispersing the capital and human resources previously amassed by HiSilicon and incentivizing large Chinese companies to invest in domestic firms that are trying to plug the capability gaps Chinese industry suffers from elsewhere along the value chain'.[15]

In 2022, China was expected to produce 26 per cent of the chips it needed, according to analysts at International Business Strategies, far adrift of the 40 per cent it had targeted for 2020. But few thought the competitive threat had let up.

'The west should take these targets seriously but not literally,' said Dan Wang, a Shanghai-based analyst at Gavekal Dragonomics. 'There's no hope that China can actually accomplish them; but they are a sign of China's intent to catch up on these important technologies.'

Certainly, China had made great strides. Its semiconductor industry commanded sales of $40bn in 2020, compared to $13bn five years earlier, and had overtaken its great rival Taiwan by value.[16] And as its industry built, its large consumer base was no less alluring for the world's largest players that it might one day seek to dispense with.

Qualcomm, the market leader in wireless chips that relied on China for two-thirds of sales, warned in its annual report that 'due to various factors, including pressure, encouragement or incentives from, or policies of, the Chinese government', some of its customers in the territory 'have developed, and others may in the future develop, their own integrated circuit products' and may use them 'rather than our products'.[17]

China was forecast to build 31 major semiconductor factories in the four years to 2024, ahead of 19 opening for business in Taiwan and 12 in the US, according to industry group SEMI.[18] Several of those would be built by Semiconductor Manufacturing International Corporation (SMIC), which in February 2022 said that local manufacturers wanted to source chips domestically because they feared other governments would prioritise their own users. In a glimpse of the future, the company installed 'non-A lines' at its factories – production lines that were completely free of American chipmaking equipment.

There was ample investment being made away from manufacturing too. Just as Apple, Amazon and Meta had pockets deep enough to design their own chips, so did the Chinese platform companies Baidu, Alibaba and Tencent. Aside from its internet giants, China had designated several thousand 'little giants', a new generation of highly specialised start-ups designed to develop the core technologies it lacked.

A sign of the money-no-object approach the Chinese leadership was taking was much in evidence at one of the sector's flagship companies. Tsinghua Unigroup contained ventures part-owned by IT company Hewlett Packard Enterprise and Intel and had once tried to acquire the US memory-chip maker Micron. When it went bust in July 2021 after amassing $30bn of debt it was resuscitated the following April with a $9.4bn injection from a government-backed entity.

The question remained whether China could close in on Taiwan and South Korea at the advanced process nodes of chipmaking. There was a virtue in the older technologies that still commanded vast volumes and could be used to make microcontrollers and chips that managed power supplies. Mature nodes of 65nm and above still accounted for nearly 64 per cent of global chip output in 2022.[19] According to analysts at International

Business Strategies, demand for 28nm chips would more than triple to $28.1bn in the decade to 2030 and by 2025, 40 per cent of the world's capacity for that process node would be based in China.[20]

However, the US was particularly focused on stopping China's progress at 14nm and lower, judged to be the process technologies key to furthering high-performance computing that powered artificial intelligence – but also the latest crop of smartphones. China was intent on continuing its advance. One notable revelation came in July 2022 when it emerged that SMIC had succeeded in developing a 7nm chip, despite the chipmaker's presence on the US entity list and being banned from using the most advanced lithography machines. US analysts at TechInsights bought the chip, which they said was made for use in cryptocurrency mining. They judged the process it followed to be a 'close copy' of the one used by TSMC but would very likely be a stepping stone to SMIC achieving its own 7nm process. The chip's designer, MinerVa Semiconductor Corp, said on its website that mass production began in July 2021.[21]

The development suggested that the manufacturer had found a work-around and that while China couldn't muster the volumes yet, actually it could be inching towards the process nodes available in the US and Europe.

Industry veterans held out hope that China would gain greater respect for intellectual property in time. 'Early on the Japanese were using IP willy-nilly but as soon as they started to develop it themselves, they grew up and protected it for everybody,' said Sir Peter Bonfield, the TI, ICL and BT veteran who joined TSMC's board in 2002. 'I think China will go through the same thing. They are probably on a bit of a journey.'

But the US was not prepared to give it the benefit of the doubt. On 7 October 2022, it tightened its 'chip choke', with a new export ban intended to curb all aspects of China's chip development. The controls restricted China's ability 'to obtain advanced computing chips, develop and maintain supercomputers, and manufacture advanced semiconductors', which the US Bureau of Industry and Security said were used to 'produce advanced military systems including weapons of mass destruction; improve the speed and accuracy of its military decision making, planning, and logistics, as well as of its autonomous military systems; and commit human rights abuses'.[22] What is believed to have stiffened its resolve is Apple's plan to use memory chips from YMTC for some iPhones – a decision that was put on hold soon after.

This time it was personal too. US citizens and permanent residents were banned from supporting China's efforts, which led to equipment makers withdrawing their support staff from chipmaking plants 'like a Dunkirk evacuation', one worker said.[23] Key allies prominent in the chip industry, notably South Korea and the Netherlands, also came under pressure to pick sides.

If China was to achieve semiconductor self-sufficiency, it would further divide an already divided world. Imagine two computing ecosystems, two sets of devices, two halves of the world that did not talk to each other. In time, China could even impose bans of its own.

And pulling apart a global supply chain fails to understand its circular nature. The US may manage to block off China's progress at the advanced nodes, but if it cannot maintain the factories that produce older technologies either, its customers may suffer. One of China's largest customers is still the US, which is a long way from replacing itself or sourcing from elsewhere the vast quantity of cheap chips it devours every year.

The stand-off is a recipe for misunderstanding and conflict. It's hard to reproach China for trying: its leaders can clearly see the geopolitical power conveyed by industrial leadership. But its methods or, come to that, the US retaliation, are far from helpful.

Perhaps the least contentious point to make is around cost. The productivity gains made by the computing industry over the last 50 years have come from collaboration, making new advances once not twice. Regional software is the epitome of deglobalisation. A Semiconductor Industry Association and Boston Consulting Group report reckoned that fully self-sufficient regional supply chains would require at least $1tn in incremental upfront investment and result in a 35–65 per cent overall increase in semiconductor prices, ultimately driving up device costs for consumers.[24]

## Powerful Partnerships

It was, according to one technology analyst, 'the biggest return of a prodigal son since Steve Jobs went back to Apple'.[25] In the quest for media cut-through, Silicon Valley watchers were not short of hyperbolic soundbites. But the announcement on 13 January 2021 that Pat Gelsinger would be the next leader of Intel deserved every breath of excitement it generated.

Since he had departed the famous chip company 12 years earlier, Gelsinger had often been spoken of as the prince across the water. He was the leader the toilers at Intel's Santa Clara headquarters dreamed of returning to right a decade of wrongs.

By 2021, Intel was listing badly. Fresh delays to its next generation of chips were letting rivals steal a march. An already aggressive corporate culture had grown sharper still. An activist investor called for change. And even Intel's cherished, by this stage almost unique, business model – of both designing its own chips and manufacturing them – was showing signs of strain.

The leadership had drifted too. Brian Krzanich, a leftfield choice to be CEO from 2013, presided over the emergence of

Intel's manufacturing problems and left under a cloud after a past relationship with another employee emerged. Bob Swan, the finance chief who stepped in, was at best a stopgap.

In contrast, Gelsinger was a 30-year veteran who knew the company inside out. Even his rural backstory bore similarities to that of Robert Noyce. Brought up on a farm in the Amish part of rural Pennsylvania, Gelsinger was spotted by an Intel recruiter after excelling at his state technical institute and, at the age of 18, took his first plane trip to begin work as a quality-control technician in Silicon Valley.

He rose to become chief architect of the pivotal 486 microprocessor and Intel's first chief technical officer, but always wanted more. During a minor career lull, at the age of 31 Gelsinger wrote out a personal mission statement that set out his ambition to one day become Intel chief executive, but after being passed over for Paul Otellini, in 2009 he left for EMC, the computer storage giant.

———————

His triumphant return sent Intel shares up nearly 8 per cent on the day. Gelsinger reached into the past as he sent a note to new colleagues, invoking the trinity of Noyce, Grove and Gordon Moore.

'My experience at Intel has shaped my entire career, and I am forever grateful to this company,' he wrote. 'To come back "home" to Intel in the role of CEO during what is such a critical time for innovation . . . will be the greatest honor of my career.'[26]

Gelsinger's Twitter feed betrayed his folksy sensibilities. Corporate announcements were mixed with pictures of his mother and grandchildren and choice Bible verses. Two days before he gave notice he would steer Intel on a new course, Gelsinger tweeted Psalm 33:11, a line that invokes determination and steadfastness. 'But the plans of the Lord stand firm forever, the purposes of his heart through all generations.'

Just 36 days into the job, on 23 March 2021 Gelsinger set out his stall with three major pieces of news as his hands flailed around enthusiastically on a video presentation entitled 'Intel Unleashed'. The production problems encountered with Intel's 7nm chip process had been fixed, he insisted, and Intel would pour billions into building new factories to catch up on the competition. Saving the biggest news for last, Gelsinger, in dark jacket and bright blue shirt, revealed that Intel would begin making chips to design specifications supplied by its customers as it adopted the 'foundry' model pioneered by TSMC 34 years earlier.

'The digitisation of every industry is accelerating the global demand for semi-conductors at a torrid pace, but a key challenge is access to manufacturing capacity,' he said at the event.[27]

In a way, it was going back to its roots, manufacturing to order. After all, it was how the Busicom calculator contract that yielded the 4004 chip started out. But much time had passed, and there were immediately questions about whether the mighty Intel could become a subservient supplier after decades of calling the industry's tune. One thing that filled Intel with confidence it was on the way back to greatness was that it had secured the tools to do the job. The company would be the first in the industry to take delivery of the next piece of new machinery from ASML, which remained utterly indispensable for any firm with ambitions to make the most intricate, powerful chips.

———————

During the two-year period blighted with Covid-19 lockdowns, every crane in the Netherlands was rented out and put to work on ASML's fast-growing site that dominated the town of Veldhoven. By the end of 2021 its campus, which looked north to Eindhoven and the Philips research base where the first concepts for an optical lithography machine were developed, comprised 187,000 square metres (over 2 million square feet) of office space, 58,000 square metres (625,000 square feet) of cleanroom used for manufacturing and research and development (R&D) activities, and 53,000 square metres (570,000 square feet) of warehouses.[28]

The company had almost doubled its staff numbers in five years to 32,000, with more than half of those based in Veldhoven. ASML sat at the heart of the Eindhoven Brainport region alongside chipmaker NXP, LED lighting specialist Signify and the semiconductor subcontractor VDL, plus start-ups and academic institutions, where in total almost a quarter of all Dutch R&D was spent.

Outside its experience centre, right next to an airy canteen that was built on the site of a car park, ASML proudly displayed a wall featuring many of the 15,000 patents it held, including for robots that could be used in a vacuum, master oscillators and electrostatic clamps.

Since its extreme ultraviolet (EUV) technology went into high-volume manufacturing in 2017, ASML had pushed further ahead of the competition. Still there was no one else on the market with EUV, which, among other improvements, reduced the number of steps required in chip processing by 30 per cent. The installed base of these machines had produced more than 59 million completed silicon wafers by the end of 2021. That year, despite the challenges brought about by the pandemic, ASML sold 42 EUV systems and forecast output of 55 machines for 2022. Some 90 per cent of sales came from Asia.

Had the investors that backed the 2012 fundraising held on, they would have seen their shares in ASML rise more than tenfold. It became Europe's second-most valuable publicly listed company after LVMH, the luxury goods house that included storied brands such as Christian Dior, Fendi and Bulgari.

Peter Wennink, ASML's chief executive, said he expected by 2025 that two-thirds of ASML's system sales would be EUV and the rest would come from the older deep ultraviolet (DUV) technology, measurement and inspection. It was a smaller share of EUV than he had predicted in 2018, but only because all the other business lines were growing faster than expected. Even advanced chips did not require EUV throughout. Many layers could be etched using DUV instead.

But the innovation that Intel's Pat Gelsinger was excited by promised to push the boundaries of lithography further still. ASML's latest prototype, created in early 2022 and due to go into high-volume production by 2025 or 2026, was for a high-NA (numerical aperture) EUV tool. By adding larger mirrors, a thin film of water and tweaking the entrance angle of the light, the light wavelength could be shortened from 13.5nm to 8nm. With larger optics, the hope was that the machine could print smaller features with higher density, reducing patterning costs for customers once again. That would offset the machine's sticker price, which was rumoured to be almost double that of the first EUV tools.

In a speech given in September 2022 at the opening of the Eindhoven University of Technology's new academic year, Wennink tried to pull together an industry that risked pulling itself apart.

'Breakthroughs in the chip industry are the result of the systemic integration of knowledge and competences across a seamless global network,' he said. 'That network is built on technological extremes, mastered by only a handful of companies. To work towards technological sovereignty, we need to create a global network of mutual dependencies based on strong local relevance.' He went on to say that 'powerful partnerships are not founded on power, but on capability, trust, transparency, reliability and a fair sharing of risks and rewards'.[29]

ASML was still barred from selling its best machines to China but that did not stop China attempting to replicate them. In February 2022, ASML said that a Chinese company, Dongfang Jingyuan Electron, was marketing products that could infringe on its intellectual property. In its annual report it noted that the company was associated with XTAL, a firm against which it had obtained $845m in damages in a US court for trade secrets misappropriation in 2019. Dongfang was a designated 'little giant', having been selected by the Chinese government as a start-up to compete with the best in the world.

## Moore No More

The geopolitical angst that overlaid the microchip industry was replicated by technical concerns. Process advances had never been easy but now, with components just a handful of atoms wide, they were getting close to impossible.

Firstly, the tiny copper wires that transferred electrical signals from one transistor to the next were so small it became difficult to move current down them. Any time delay caused by the resistance of these interconnects hit clock speed, slowing the pace at which a microprocessor carried out instructions.

Then, more devices crammed into a smaller space meant more heat was generated. This was a sign that Dennard scaling was breaking down. The idea that as transistors got smaller the power they consumed remained the same dated from a 1974 paper written by Robert Dennard, an IBM researcher. Its failure posed designers with the challenge of how to come up with chips that could power down some parts when they weren't in use to prevent overheating.

Another problem was the leakage of electric current, which increased as voltage was reduced, which could drain batteries. In all, the power and performance benefits that Moore's Law had always delivered were not as clear-cut as they used to be.

Nevertheless, TSMC planned to begin manufacturing chipsets using a 2nm process technology in late 2025 and had a target for delivering its first batch in early 2026. Samsung, which began initial production of its 3nm process node in June 2022, talked of mass production for 2nm in 2025. And Intel thought it could begin producing something similar in late 2024, which, if it succeeded, would signal it had overhauled its two greatest competitors – as CEO Pat Gelsinger had made his mission. The challenge for all was not merely production, but reliable production – improving the yield to flatten the price and give security of supply. One unspoken threat was that if the established manufacturers couldn't push forward, China had a better chance to catch up.

Intel had ambitions to combine one trillion transistors in a single package by 2030. In preparation, it even started to measure its product in angstroms, equal to a tenth of a nanometre or one hundred-millionth of a centimetre.

The company thought its version of a Gate All Around (GAA) transistor would combat current leakage because the gate would encase the entire channel. Beyond that there were 'nanosheets' as thin as a single atom that would stack horizontally to surround the channel in all directions.

Experts believed this type of innovation might contribute to an increasing number of transistors on a chip for some years, but the associated improvement in processing power was less clear. However, nothing was ruled out to meet the demand – nay, expectation – of greater processing power to propel forward artificial intelligence, climate research, aerodynamics and computational medicine. By 2025, the world's data was forecast to reach 175 zettabytes annually, which stored on DVDs would stack high enough to circle the earth 222 times.[30]

After 60 years of faithful service, silicon could be retired if alternative materials such as indium antimonide, gallium arsenide and carbon in various forms are found to offer higher switching speeds at lower power. They may be more expensive, though. Even germanium, Jack Kilby's semiconductor of choice, is being revisited decades on.

Gordon Moore, long retired to Hawaii's Big Island, doubted his 'wild extrapolation' would go on forever. In a 2015 interview to mark the 50th anniversary of his prediction, he pointed to the finite velocity of light and the atomic nature of materials as reasons why Moore's Law would probably die in a decade or so. Vast costs were another negative. But the achievement of putting several billion transistors on an integrated circuit meant there was still scope for improvements to be made. 'The room this allows for creativity is phenomenal,' Moore said.[31]

Skyscraper chips assembled 100-layer stacks of circuitry multiple times. Four stacks were possible as this book went to press, with plans to get to eight. 'You start a new ground floor on top of the top floor of the previous building,' said one industry executive.

As well as stacking, much innovation has gone into packaging, once the most basic stage of production, where, notably, China is well advanced. 'Flip chip' techniques, where the chip has a tiny blob of solder applied so it can be flipped over and mounted on external circuitry to create a bigger module, have their roots in the 1960s. Another method is using chiplets, where the semiconductor die is broken into smaller pieces, fabricated and then recombined at the packaging stage. This supposedly accelerates production time, improves yield and could lower cost.

Chief among future hopes to further the drumbeat of progress is photonic processing, which offers the promise of accelerating calculation speeds and lowering power demands by replacing electrical transistors with optical light pulses. Then there is biocomputing, which harnesses tiny organic molecules, such as proteins or DNA, to make complex calculations. And the field of quantum processing uses qubits, the equivalent of bits in traditional computing, that can possess the value of both one and zero at the same time. Beyond Moore's Law, they offer exponential growth, but operate best at sub-zero temperatures.

## Accelerating into a Slowdown

Precisely one week after Nancy Pelosi touched down at Songshan airport in Taipei for a historic visit that rattled global financial markets and stung the Chinese leadership, a who's who of the US microchip industry gathered on the south lawn of the White House.

On the morning of 9 August 2022, Arvind Krishna, IBM's chairman and CEO, was joined by Intel's Pat Gelsinger, Lisa Su from AMD and Tom Caulfield, the CEO of GlobalFoundries, to welcome the biggest state support package for their industry for decades.

The CHIPS and Science Act – CHIPS stood for Creating Helpful Incentives to Produce Semiconductors – stemmed directly from President Joe Biden's executive order to address the supply chain crisis the previous year. It provided $53bn for American semiconductor research, development, manufacturing and training, plus a 25 per cent tax credit for capital expenditure on plant and machinery.

'America invented the semiconductor,' said Biden, dressed in his regulation blue suit and sunglasses, 'and this law brings it back home. It's in our economic interest and it's in our national security interest to do so.'

He recalled a trip earlier in the year to the Alabama factory of the defence contractor Lockheed, where Javelin shoulder-fired missiles were manufactured. Each system featured 200 chips and could automatically guide itself at its target after launch, giving the gunner more time to take cover.

The US needed chips for those missiles 'but also for weapons systems of the future that are going to be even more reliant on advanced chips', added Biden, as he prepared to sign the bill into law. 'Unfortunately, we produce zero percent of these advanced chips now. And China is trying to move way ahead of us in manufacturing these sophisticated chips as well.'[32]

The Republicans who opposed the initiative said it offered a 'blank check' to companies. But there were strings attached. Any US firm benefiting from the support measures was barred for the next 10 years from expanding in China its production of chips at anything more advanced than the 28nm process.

Another initiative, America's Frontier Fund (AFF), was launched in June 2022 to steer public and private money into critical technologies that would ensure US competitiveness. It was led by Gilman Louie, an executive best known for setting up the CIA's venture capital arm, who warned in testimony to the US Senate's finance committee that another key chip supplier, Samsung, operated factories in South Korea that were 'within North Korea artillery and missile range'.[33]

The AFF crossed the political divide, uniting Eric Schmidt, the former Google chief who was a prominent Democrat fundraiser, and Peter Thiel, the Donald Trump supporter who co-founded PayPal and set up big-data firm Palantir Technologies.

Another plan was struggling to gain traction. The US government's Chip 4 alliance with Japan, South Korea and Taiwan to secure the semiconductor supply chain, co-ordinate subsidies policies and research and development (R&D) foundered because parties did not want to share trade secrets and were worried about the Chinese response.

An alternative plan was rumoured. If the 25-per-cent-by-value threshold for US export controls was not tough enough, perhaps a cumulative threshold extended across allies including the Netherlands, South Korea, Japan and the UK, under a G7 or NATO umbrella, would be.

'The US would have to do a tremendous amount of lobbying to get these other countries to align with them,' said one executive. 'But if that was something the countries wanted to do, you could really build a wall around China.'

At least the CHIPS Act was having an impact. In January 2022, Intel had committed to build what Biden described as the 'field of dreams' on a thousand acres outside Columbus, Ohio: two new chip factories at a cost of more than $20bn that were due onstream in 2025, with capacity to add six more. It followed hot on the heels of a similar-sized investment in Arizona, announced by Intel the previous year. Reports suggested the company would receive $2bn in incentives to open in Ohio, its first new manufacturing location in 40 years.

------

The US was not alone in spending heavily to build its tech sovereignty. On 15 September 2021, Ursula von der Leyen, the European Commission President, underlined intent in her state of the union address, announcing plans to bring forward a new European Chips Act that would link research, design and testing capacities across the bloc, ensure security of supply and develop new markets.

'Yes, this is a daunting task,' she said in a speech given at the European Parliament in Strasbourg. 'And I know that some claim it cannot be done. But they said the same thing about Galileo 20 years ago. And look what happened. We got our act together. Today European satellites provide the navigation system for more than two billion smartphones worldwide. We are world leaders. So let's be bold again, this time with semiconductors.'[34]

The Commission estimated it needed to put as much as an extra €125bn into digital technologies every year, of which €17bn was for semiconductors. The target was that by 2030 the European Union should account for at least a fifth of the world market by value for cutting-edge and sustainable semiconductors, up from a tenth.

Again, Intel obliged. The announcement on 15 March 2022 was music to Von der Leyen's ears. Intel said it would develop two semiconductor fabs in Magdeburg, Germany, with construction expected to start in the first half of 2023 and production planned for 2027. With initial investment of €17bn, the project would create 3,000 permanent high-tech jobs, plus tens of thousands of additional jobs at suppliers and partners.[35] To be known as Silicon Junction, Intel was reportedly subsidised to the tune of €6.8bn. But Chipzilla also increased its commitments in Ireland, Italy, France and Poland, along the entire semiconductor value chain from R&D to packaging.

The frenzy of building work didn't stop there. In November 2021, Samsung had pledged to build a $17bn fab just outside Austin, Texas, its biggest US investment, that was expected to be operational in the second half of 2024.[36] In the same year, TSMC expected to start volume production from its new $12bn factory in Phoenix, Arizona, built at the urging of the US government and capable of operating at the 5nm process level. That investment mushroomed to $40bn in December 2022 when the company confirmed a second fab would be added by 2026, operating at 3nm. Even Japan, which had clung on to some market leadership in supercomputers and robotics but whose semiconductor expertise had withered since the 1980s, made the sector a national priority on a par with food and energy.

More impressive was South Korea's vast $451bn, decade-long investment programme, led by Samsung and its closest domestic competitor, SK Hynix. Supported by generous incentives, the government planned to create a 'K-Semiconductor Belt' in the western part of the country that would cluster companies involved in chip manufacturing, materials, equipment, and design. President Moon Jae-in said the plan was to make 'pre-emptive investments that will not be shaken by external shocks' so the nation could lead the global supply chain.[37]

It was Rock's Law writ large. Arthur Rock, the industry's legendary early financier and one-time Intel chairman and Apple director, posited that the cost of a semiconductor plant would double every four years. And so large were the design costs for the next generation of transistors – the building blocks of integrated circuits – that all but the highest-volume chips risked being rendered uncommercial. Analysis by International Business Strategies (IBS) reckoned the cost of developing a mainstream 3nm chip design was an eye-watering $590m, compared to $416m for a 5nm chip.[38]

All told, the industry's capital expenditure was on track in 2022 to nearly double from the $150bn spent globally in 2021. There was much more to come. Five major territories – China, the US, the EU, Japan and India – collectively committed subsidies of $190bn over a decade.[39] Before most of those announcements, from 2020 to 2030 global manufacturing capacity was predicted to increase by 56 per cent.[40]

TSMC's Morris Chang was doubtful all of these political initiatives would have the desired effect, calling America's attempt to grow domestic chip production as 'a wasteful, expensive exercise in futility', because the nation lacked the specialised talent pool. Then again, the industry suffered from an acute shortage of highly skilled workers everywhere. Chang's contention that manufacturing chips in the US was 50 per cent more expensive than in Taiwan raised questions about competitiveness that was bound to have a knock-on effect on product costs.[41]

Meanwhile, the economic outlook for microchips was weakening. Optimistic semiconductor executives supported the notion that the rate of industry growth would step up over the next decade compared to the average 5 per cent growth of the previous decade as chips decoupled from consumer demand to be embedded in more industrial applications and more were required per device. For investing into a downturn, they would eventually be rewarded. But the pessimists couldn't shake the feeling that bust would once again follow industry boom and the prices they could charge for chips would crater. Much could change in the three years it took to build and fit out a new fab.

The sheen had certainly come off Gelsinger's leadership of Intel. Eighteen months into his reign, the company's shares had slumped by almost a third and its market value had been overtaken by old foe AMD for the first time. On 28 July 2022, Intel published earnings well below Wall Street expectations as PC sales slowed and, surprisingly, demand from data-centre customers softened too. 'This quarter's results were below the standards we have set for the company and our shareholders,' Gelsinger said. 'We must and will do better.'[42] On 22 February 2023, Intel slashed its dividend to the lowest level for 16 years amid slowing demand.

Compared to the 26.2 per cent spike in 2021, in August 2022 the official industry tracker World Semiconductor Trade Statistics lowered its outlook to 13.9 per cent growth for the current year, down from an earlier 16.3 per cent prediction. It pencilled in just 4.6 per cent growth for 2023, the lowest since the US–China trade war bit in 2019. By November 2022, it had cut its growth outlook again, to 4.4 per cent for 2022 and a 4.1 per cent contraction in 2023.[43] Samsung got 2023 off to a poor start, reporting fourth-quarter operating profit at an eight-year low as the slowing global economy hammered demand for chips and its electronic goods. TSMC cut its capital expenditure.

One thing was for sure: demand might weaken but the great expansion in supply meant the world would never run short of microchips again. What was far from clear was the security of that supply, and the price that could be charged for it.

# PART FOUR

## ARM (2022-)

# Chapter 16

# ALWAYS ON

## Back Home

The collection of golden tetrahedron structures artfully hung over the reception at Arm's new Cambridge headquarters was just a taste of the decor that ran throughout. Inspired by the sharp angles of crystalline silicon, the jagged pattern extended to the walls and ceiling of the airy atrium, dividers that stood between workspaces, triangles of frosted glass, the grey, vertical fins that trimmed the exterior, and even the fancy doorhandles.

One day in July 2022, staff in shorts and sporting beards relaxed over coffee in twos and threes, peering at their smartphones and chatting. This base, built on the Peterhouse Technology Park initially to house 1,700 Arm employees, might have offered a future vision of the workplace, but it could not shake off echoes of the past.

Arm was still here, on Fulbourn Road in the district of Cherry Hinton, where it returned from Harvey's Barn in Swaffham Bulbeck to displace its former parent company Acorn in 1994 and had been incubated a decade earlier. The year of the move was the year of Nokia, a piece of business won via its licensee Texas Instruments that set Arm on a trajectory and made it a company that mattered in the technology universe. Remarkably, almost three decades later, as it tailored its power-efficient designs to new markets with higher performance demands, it still did.

Today, the waterworks had passed back to the local water company and the Silver Building, built when Acorn was flush with confidence and money, had been demolished to make way for a car park, which was decorated with more triangles on its exterior. From there visitors emerged, crossing an avenue of lime trees, to enter the new building that was constructed on more farmland to the rear of the original site.

The building next door, opened in 2000 by Stephen Byers, the trade and industry secretary, and where staff assembled to celebrate the SoftBank deal with a beaming Masayoshi Son in July 2016, was still in use but the executives were long gone. On the balcony, Simon Segars' old office was now a 'retreat room' and there was a blue plaque to mark the workspace of Mike Muller, one of Arm's original 12 engineers and chief technology officer who had been the key contact for the vital customer Apple before his retirement in 2019.

Compared with the futuristic statement campuses built by the US tech giants, Arm's home remained modest, looking out across the road on clusters of semidetached houses with curtains drawn to protect against the harsh sunshine. And it was rarely full, given working from home still held sway at the tail-end of the Covid-19 lockdown.

Modern, global companies try not to get pinned down by place, particularly when intellectual property can zip around the world at the touch of a button, with business-class executives flying not far behind it. But it was impossible not to observe that Arm's power base had been drifting to the US ever since the Artisan acquisition in 2004. Now, most of Arm's leadership team was based in the US, where its main site was on Rose Orchard Way in San Jose, California. Of the 13-strong executive team, six were British and five were based in the UK with the rest in the US.

Still, staff gathered in numbers in the Cambridge atrium for the big moments in Arm's history, such as, in February 2022, only the third handover from one leader to the next, from Segars to Rene Haas.

---

Rene Haas had a long history working in semiconductors but for Arm he was a departure. He was the first chief executive not to have spent his formative years in Cambridge and Harvey's Barn, but given the skew towards Silicon Valley it was only a matter of time before Arm appointed an American boss. Haas was well known to the company beforehand, having spent seven years at Nvidia, where he led its division selling into notebook computers and promoted the Tegra chip that used Arm designs. When Segars became chief executive in 2013, the expansive Los Angeles Lakers basketball fan was his first senior hire and became a close lieutenant.

To get there, Haas had served a long tour of duty, compared to Segars, effectively a one-company man. He was brought up in Rochester, an old industrial city on Lake Ontario in New York State famous as the home of Eastman Kodak and Xerox. Haas was a radio DJ in college but traced his interest in technology back to his mum, a teacher, and his physicist dad, who brought home a Texas Instruments Silent 700 computer terminal for his 11-year-old son. It featured a built-in modem that could dial into the local Xerox mainframe. 'I hooked it up and next thing you know I'm typing into a machine that's typing back and I was sold on computers,' Haas said.

He received his Bachelor of Science in electrical and computer engineering across the state in Potsdam's Clarkson

University in 1984 and held engineering positions at Xerox and TI before spending a decade at NEC. There, he began as a field applications engineer, supporting customers using MIPS-based processors and other chips, before moving up to become a sales director looking after key accounts including HP, Compaq, Intel and Cisco.

It was a period when several computer architectures were still vying for leadership and MIPS possessed more momentum than the juvenile Arm. By 1999, when Haas became regional sales head at Tensilica, a two-year-old company whose technology contributed to application-specific processor cores, Arm showed signs that it was pulling ahead.

There was irony here. Arm's early engineers had been envious of MIPS at the outset, but the firm lost its way trying to become a computer vendor. Tensilica was set up by Chris Rowen, one of MIPS' founders, and hired several of its staff. 'We looked at Arm as the gold standard because it had just gone public,' Haas said. 'That was the company we compared ourselves to.'

When he joined Arm in October 2013 it was as vice president of strategic alliances, soon stepping up to the executive committee as chief commercial officer in charge of sales and marketing. Haas was based in Shanghai for a time, and, as president of Arm's IP Products Group (IPG) from January 2017, he put some focus on diversifying the product portfolio so that it targeted key markets and increased investment in each of them.

When Nvidia and Arm partnered on 'deep learning inferencing' in March 2018 – essentially helping to train computers to think like a human brain – Haas was front and centre. At Arm's 2018 TechCon event he launched the Neoverse platform for data centres, 'a cloud-to-edge infrastructure'. The trend for keeping data local to customers to improve response times and save bandwidth, instead of holding everything in a giant, central cloud – so-called 'distributed computing' – meant there was demand for more, and smaller, data centres and therefore more chips. It was good news for Arm.

Underscoring his importance to the firm, Haas was deployed the day after the Nvidia deal was announced to try to reassure customers. After all, he had seen both sides at work. But his biggest challenge beginning in February 2022 was to steady the company after the $60bn sale to Nvidia collapsed, especially as Segars had warned that an initial public offering (IPO) would pile on pressure to deliver short-term profit and sales growth and 'suffocate our ability to invest, expand, move fast and innovate'.

Haas, tall with salt-and-pepper hair, prominent nose, grizzled and toothy, was quick to step away from his former colleague Segars' messaging. 'There isn't anything Nvidia and Arm could do together that we can't do by ourselves,' he told media the day after his appointment.

But because Son's Plan B was to list shares in the business, Haas had to get it into better financial shape. One month later, Arm said it planned to cut up to 15 per cent of its workforce, a cull that could have approached 1,000 people, with losses mainly in the UK and US.

It was only six months since SoftBank confirmed that it had fulfilled its post-takeover undertakings. Five years on from its purchase of Arm, the Japanese investor had maintained the company's global headquarters in Cambridge, more than doubled UK headcount to 3,560 and ensured that the majority of those workers filled technical roles.

With what seemed like indecent haste, now it was unpicking that growth. Amid grumbles about staff morale, headcount was reduced by 1,250, marking an 18 per cent decline overall – although about half of these roles were with the underwhelming internet of things (IOT) software platform, Pelion, and associated assets that were being hived off and remaining with SoftBank. Many of the rest were administrative and support staff.

But the company was also hiring, as Haas looked to expand Arm into new fields. 'We have a first-world problem in that we can almost play in any market,' he said. 'So we have to make really wise choices to make sure that the investments we're making are going to give you the best return.' The chief executive had been around; now it was Arm's chance to do the same.

## Tug of War

Compared to its speedy and somewhat subdued departure from the public markets in 2016, the campaign for Arm's return to the London Stock Exchange was long and loud. Stung by criticism from investors that, post-Brexit, Europe's financial capital had lost its edge versus the go-go New York and Shanghai markets – and perhaps even its close neighbour Paris – ministers made strenuous overtures to SoftBank to lure back the nation's flagship technology company.

Arm, with its global reach, decades-long track record, operating at the cutting edge of computing, with a customer base to die for and a headquarters just 60 miles away, was a 'must-have' to show London could compete. Six years after its buyout offered a sign that global investors still wanted to sink their cash into the UK, British politicians and financiers were intent on welcoming Arm again for precisely the same reason.

It was an uphill battle. Son had already declared that instead of London he favoured New York, where most of SoftBank's companies had floated their shares in pursuit of lofty valuations. 'We think that the Nasdaq stock exchange in the US, which is at the centre of global hi-tech, would be most suitable,' he said.[1]

Lord Grimstone of Boscobel, the UK government's craggy-faced investment minister, led the charge. On a visit to Japan in June 2022, the former Treasury mandarin who had masterminded numerous privatisations for Margaret Thatcher's government and then become vice chairman at investment bank Schroders, put the UK's case to Son.

Having relaxed stock-market rules to lure technology companies in particular, Grimstone proposed to SoftBank something distinctly unusual: a dual primary listing. The structure would allow Arm full access to the pools of liquidity in New York and London and membership of the Nasdaq Composite index, which comprised more than 2,500 technology stocks, plus entry to London's flagship FTSE 100, featuring the largest 100 publicly listed stocks that qualified.

There was another dimension to the charm offensive, which the prime minister Boris Johnson joined by writing a letter to SoftBank executives that extolled the capital's virtues. Sir Alex Younger, the former head of the British secret service, MI6, said the government should 'strain every sinew' to keep Arm in the UK, saying that the nation's 'future security depends on our ability to sustain and grow a strong science and technology base'. He added: 'There is a direct security and military aspect to this. But more important is the economic strength it generates.'[2]

All of this meant it was a blow when, three weeks after the Grimstone meeting, Son assured shareholders at the company's annual general meeting at the Tokyo Portcity tower that 'Nasdaq is the favourite'. The billionaire added: 'Most of Arm's clients are based in Silicon Valley and stock markets in the US would love to have Arm.' That was despite 'a strong love call' from London and bankers still assessing where regulations were most suitable.[3]

What set back the UK campaign even further was the political crisis that came to a head on 7 July 2022 when Boris Johnson resigned after losing the confidence of his party following a string of scandals. Grimstone and fellow ministers departed too, so the conversation with Arm ground to a halt. There were efforts to rekindle the charm offensive under Johnson's short-lived successor, Liz Truss, and again under Rishi Sunak, who hosted Haas at Downing Street before Christmas 2022.

It was a hard sell. A dual listing undoubtedly came with additional cost and complexity, and potentially required two sets of accounts to meet differing accounting standards, plus two governance codes that dictated how Arm's board should be run. Haas, supported by Son via video call, tried to convince Sunak that what really mattered to the UK was Arm's ongoing commitment to expand and hire where the majority of its workers were already based. And that could be made easier if the government simplified visa rules so that engineers could be brought in from all over Europe with less hassle. There was an air of inevitability when Arm eventually declared for New York on 3 March, 2023. After engaging with the government over several months, Haas said that SoftBank and Arm 'have determined that pursuing a US only listing of Arm in 2023 is the best path forward for the company and its stakeholders'. It sought to soften the blow to British pride with plans to open a new site in Bristol, grow headcount and 'maintain its headquarters, operations and material IP' in the UK, as well as 'consider a subsequent UK listing in due course.'

---

Meanwhile, Arm's financial performance was improving. The company reported record revenue in the quarter to June 2022 of $719m, up 6 per cent on a year earlier, and profits rebuilt to $414m, up 31 per cent on the same period, even before the benefits of its reduced cost base fed through. Following some licence delays, sales and earnings fell back in the following quarter, but leapt in the final three months of 2022, to $746m and $450m respectively.[4]

It was noticeable that royalties were mushrooming. In a fast-moving industry, Arm's business model was a slow burn. New pieces of intellectual property took 18 months to develop and then a customer might take 18 months to design them into a chip. By the time the product was selling, four or five years could have passed before royalties came in.

One reason for the upward trend was that after more than a decade of trying, Arm was finally getting traction in the data-centre market. Market tracker Omdia said its market share in the second quarter of 2022 rose to 7.1 per cent, its highest to date.[5] Arm-based chips that delivered higher computing power but ate up less energy were easy to customise and ideal for smaller data centres on the cloud's 'edge'. One research report, from Canalys, said Arm could grab 50 per cent of the market for cloud comput-ing processors by 2026, a striking jump.[6]

Much of that was down to Amazon Web Services and its Gravi-ton chip, but Huawei and others had also turned to Arm. Microsoft and Google began using Arm-based chips made by the expanding Ampere Computing, which already supplied Oracle and data-cen-tre specialist Equinix. Nvidia followed, with its Grace processor. It was a flurry of activity ideally timed for Arm, just as smart-phone volumes tipped into decline. Reaching back to its earliest years, that expertise from designing for battery-powered handsets remained exceedingly relevant in these newer markets. What was different was that chips might be the most expensive component in a smartphone, but server processors cost far more. The royalty rates the company could charge were much higher.

Canalys also predicted that by 2026 Arm would have grabbed 30 per cent of the PC market, which was still dominated by its old foe Intel's x86 processor. And then there was automotive, where Arm reported that its growth in powering in-car entertainment systems and advanced driver assistance systems (ADAS) had outpaced both smartphones and data centres as part of a market that was on course to double in five years.

'It's our time,' Haas wrote in an email to staff, confident that Arm was well positioned in growth markets compared with Intel's x86 and the less proven RISC-V. It helped that electric cars contained twice the number of chips as their petrol-powered cousins and the semiconductors were second only to the battery in terms of component cost. Some carmakers were still suffering from supply shortages too.[7]

It also helped that, under SoftBank ownership, Arm had gone from designing general-purpose CPUs, which were used across smartphones and data centres, to tailored products that no longer had to compromise in certain categories. Those decisions, taken five years back, might even help Son get a decent price when he eventually came to realise his investment.

———————

Arm also appeared to have the embarrassing situation in China under control. In April 2022, Shenzhen officials finally approved the removal of Arm China's rogue chief executive, Allen Wu, striking his name from the firm's business records and creating a new 'chop' – the all-important wooden stamp – to authorise official documents.

The pursuit of Wu through the Chinese legal system had gone nowhere. It was only when Arm moved to transfer its shares in the venture over to SoftBank, that the government acted. To end a two-year standoff, police turned up at Arm's 24th-floor office in Shenzhen and took away two of Wu's security guards.[8]

Wu wasn't done, advising colleagues to work from home until further notice. He told one Chinese media outlet that 'SoftBank released false news that it was able to reach a deal with certain officials . . . this is a clear violation of the law!'[9] Loyalists blocked messages to Arm China's 800 staff from the Arm-backed shareholders. Eventually, control of the venture's IT system, social media accounts and bank access was taken back and two co-CEOs were installed, one of whom, Renchen Liu, a Chinese government adviser and compromise candidate, was accepted as the company's legal representative and general manager.

'Arm remains committed to the Chinese market, Arm China as an independent operating company majority owned by Chinese investors, and to the success of Arm's Chinese ecosystem partners,' a SoftBank statement read.[10] Selling its designs into China through the joint venture normalised, but the joint venture's efforts to create China-specific designs were less popular.

Where uncertainty lay was in the latest US export controls. The focus on stopping the most advanced equipment and design tools from reaching China had long since expanded to reach the less visible, but still crucial, computer architectures. New performance thresholds that covered any technology enabling artificial intelligence hit Nvidia specifically, but also Intel. Arm's building blocks were not affected on their own but the new rules were so loose that if they were considered in combination with other pieces of technology they possibly could be.

The open-source RISC-V offered an alternative, but far better to maintain access to the dominant Arm architecture. That could explain why China appeared reluctant for Arm to distance itself from Arm China. The transfer of its stake to a new SoftBank entity was regarded as a precursor to the group's long-awaited IPO, but getting the paperwork processed proved problematic.

---

Not everything else was rosy either. In August 2022 Arm sued Qualcomm, one of its largest customers, alleging the breach of licence agreements and trademark infringement. Arm demanded the destruction of some designs it said were transferred without its consent from Nuvia, a start-up Qualcomm acquired a year earlier.

It was a classic semiconductor industry spat, with parallels of the row between the interdependent giants Apple and Samsung that rumbled on for years, only to be resolved in 2018. However, it was rare for Arm to be drawn into something where the stakes were so high.

Nuvia was established by a group of former Apple chip designers to make processors for computer servers and Qualcomm wanted it to boost its own customised processor cores for laptops too. Qualcomm and Apple were already at loggerheads, as the latter raced to reduce its dependence on the former by designing its own phone components. More interestingly, Gerard Williams, Nuvia's leader, had spent 12 years at Arm's design centre in Austin, Texas, before being hired by Apple as it scaled up its in-house design capabilities. Williams had been hugely influential implementing Arm's 64-bit architecture for Apple, which enabled it to open a lead over smartphone rivals in 2013.[11]

Qualcomm believed it had licences covering the Nuvia technologies and said the lawsuit marked 'an unfortunate departure' from the pair's longstanding relationship. It was a reminder that great technology came down to great people; not just those that invented it, but those that put it to work too. Even if Arm lost its best people, as it did from time to time, they took their Arm expertise elsewhere and that meant the ecosystem kept growing. Among the army of 13 million developers that worked with its architecture, Williams was one of the most valuable advocates.

Scheduled to begin in September 2024, the lawsuit was a risky strategy. Qualcomm was an industry leader, whose efforts were expected to increase Arm's market share in PCs and cars. Analysts speculated that Arm really wanted to negotiate a higher royalty rate rather than force Qualcomm to destroy Nuvia's designs. Either way, the case could play into the hands of RISC-V, the upstart royalty-free alternative, or generate uncertainty just as Arm was trying to find new investors.

———————

Since agreeing Arm's takeover with Nvidia in September 2020, SoftBank's financial woes had not eased. In August 2022, the company reported a record £19bn quarterly loss after a global sell-off in tech stocks hit the value of its portfolio. SoftBank sold its final stake in Uber, the ride-hailing app, disposed of more shares in Alibaba and resolved to cut costs. 'I am ashamed of myself for being so elated by big profits in the past,' said Son.[12]

The work towards listing Arm's shares on Nasdaq continued. During autumn 2022, Arm remade its board with Wall Street squarely in mind, recruiting former executives from AOL, Intel and Qualcomm. Most eye-catching was the appointment of Tony Fadell, the engineer who secured Arm's second wave of success by bringing it into Apple's category-defining family of devices starting with the iPod in 2001, and then campaigned to keep it there when Steve Jobs' eye was drawn by Intel. Fadell described Arm as 'silicon's lingua franca' and said he would 'help ensure every future builder is enabled by this essential company'.

However, the weak market conditions showed no sign of letting up, as the US Federal Reserve continued increasing interest rates to curb rampant inflation and fears of a global recession mounted. Shares in tech stocks slumped. The ambition to float Arm by March 2023 was pushed out to 'sometime in 2023' but the company insisted its plans were advanced. Son announced he would put his exclusive focus on Arm, 'the source of my energy, the source of my happiness, the source of my excitement'.

An alternative future that had clearly been given some consideration either side of the Nvidia deal was for Arm to become owned by a consortium involving many of its largest customers – rather than just one of them. That way its neutrality could be enshrined and the broader industry would have an even stronger interest in its future prosperity. Intel's Pat Gelsinger said that if such a plan for Arm was to emerge, 'we would probably be very favorable to participate in it in some manner', while Qualcomm's Cristiano Amon said, 'We're an interested party in investing.' South Korean chipmaker SK Hynix was linked to Arm too.[13, 14]

In September 2022, it was reported that Son was heading to South Korea to discuss an alliance between Samsung and Arm. A sale was immediately speculated – or a stake sale at a discounted price ahead of the 2023 IPO to give Arm a new cornerstone investor.

'When Chairman Son comes to Seoul next month, he will likely make a sort of proposal about Arm,' Samsung's vice chairman Jay Y. Lee, its de facto leader, told reporters when he landed at the South Korean capital's Gimpo Airport after returning from a two-week trip to Europe.[15] Lee was the grandson of Samsung's founder, who had announced his charge into the semiconductor industry almost four decades earlier, relying on the 'great mental fortitude and creativity' of his people.

A purchase of Arm would pose the same problems for Samsung that Nvidia encountered. But for the company that led the world in memory-chip supply it represented a unique opportunity to transform its presence in microprocessors too.

It was clear Arm's value endured. The trouble was that everyone wanted to own a piece of it – but no one could agree how.

## Milestones

With so much to occupy the company, it was easy to miss that in February 2023 it reported it had passed another substantial milestone. Some 250 billion chips based on its designs had been sold and were connecting the world, in smartphones, laptops, industrial sensors, cars and data centres. Another 1000 chips were being added to the installed base every second.[16]

The numbers were a reminder that for all the distractions, ownership wrangles, competitive threats, relentless technological change and vast investment in pursuit of new markets, Arm was everywhere, and increasingly so. Through working with more than a thousand partners, its foundational technology had achieved breathtaking range, supplying tech from the smallest sensor to giant supercomputers.

The UK was famous the world over for the Royal Family, James Bond, the BBC, Harry Potter and Premier League football. Brands such as Land Rover, Burberry raincoats and Johnnie Walker whisky were much sought after, marketed heavily and spoke of quality and heritage. But here was a little speck of Britain pulsing with data billions of times over, right across the planet, largely unheralded outside its industry.

Even though it had eyes on the future and the next application for its designs, Arm was still profiting handsomely from the past. The ARM7TDMI processor that broke through with Nokia and TI in 1997 was embedded in factory machinery, washing machines and windscreen wipers. It still shipped 200 million units in 2020 alone, extending its shelf life far beyond estimates.

In addition, almost half of the chips deployed to date – and typically three-quarters of the 30 billion annual run rate – were the low-cost and energy-efficient Arm Cortex-M 32-bit microcontrollers that had been embedded in IOT devices and cars, managing power, systems, touch-screens and batteries. That family of processors dated from 2004.

In the company's 2010 annual report that celebrated its 20th anniversary, the chief executive at the time, Warren East, marked the shipment of 25 billion chips to date and wrote that 'we are now looking forward to a future where another 100 billion Arm processor-based chips will be shipped over the next ten years'. In the event, it was far more.[17]

But the old guard was no longer counting. 'I stopped being shocked at ten billion sold,' the blueprint's co-creator Sophie Wilson said in one interview.[18] Really, with artificial intelligence, supercomputers, robotics and IOT all hungrily demanding more chips, the one trillion mark was not so far off.

When it introduced the ninth version of its architecture, ARMv9, in 2021, the third chief executive, Simon Segars, declared that it would be at the forefront of the next 300 billion Arm-based chips. The company predicted that soon all the world's shared data will be processed on Arm; either at the end-point, where it is viewed or collected via devices and sensors, stored in the computing cloud, or in the data networks that connect them.[19]

In reality, the 30-billion-a-year run rate was less relevant than the billions of lines of additional software code written annually to operate on Arm-designed hardware. So much was available in the open-source computing community that, rather than starting from scratch, developers rarely had to look anywhere else for their next widget design.

---

It is worth considering why all this matters. Arm's message to the world of putting collaboration before competition seems to come from a simpler time. By working with everyone it played a role in hastening broad technological development, empowering devices that touch everyone. Working with each new licensee, its architecture became more robust and versatile. Software volumes mounted.

And as digital giants have taken greater control over the silicon in their devices, Arm has remained on the right side of useful and economical – an effective enabler and partner. To echo some of Morris Chang's words about TSMC, market expectations of a weak competitor – if Arm generated any expectations at all on day one – eventually dissolved into the realisation that Arm could actually become a strong supplier.

Its model was not new. Technology licensing dated back at least 70 years to Mother Bell's Cookbook, when the US government was eager to accelerate the development of the transistor. As those transistors flooded the globe and rivalries flared, Arm somehow continued to stand between nations – for now.

Arm's skill was in turning the constraints of its birth to an advantage. If it had been replete with money, its rulebook would not have been so nifty. Those constraints continued so that its first managers identified all that it could not do: design whole chips, manufacture them, insert them into its own devices.

By process of elimination, Arm knew where success lay: getting the message out, finding customers from a broad array of sectors, forging a standard that industry could cluster around. It would have been naive to think that technical excellence would shine on its own. Proprietary formats such as Sony's Betamax demonstrated how easy it was to get crowded out of a fast-growing market by whoever grabbed critical mass. That is why almost the first thing that the company's founding chief executive Sir Robin Saxby did was get on a plane to tout for business in Asia.

It is easy to dismiss Arm's pivotal success with Nokia and TI as merely being in the right place at the right time. But unless it had spread the word, encouraged licensees to experiment, expressed a willingness to adapt by introducing the Thumb extension, it would not have broken through.

There are no straight lines in business. Apple's modest introduction to Arm in 1986 did not lead directly to the investment that funded the spin-out from Acorn in 1990. Nor did that residual shareholding have any bearing on the outcome when Apple went looking for a processor for its iPod in 2001. However, Apple's request for an architecture licence in 2008 stemmed straight from what it had seen of Arm over the preceding seven years, no doubt compounded by the star designer Dan Dobberpuhl's own experience working within its rules.

In a world where proprietary technology had dominated, Arm has been sufficiently relaxed to let its instruction set have life beyond the company, so that an army of developers became an army of advocates. That helped it progress as a silver thread through decades of innovations: from basic mobile phones to smartphones, eventually IOT, and now data centres and cars offer two promising new chapters.

If that sounds effortless, this book shows it hasn't been. Arm has had to take the long-term view, dealing with false starts and hoping that new markets could be cultivated before mobile demand softened. The company has at times made its own luck, particularly in the area of data centres, a 15-year odyssey where it tried to stimulate demand by investing in promising start-up firms that designed high-powered chips for computer servers. One of them was acquired by the mighty Amazon in 2015, providing momentum today.

To the UK, a country grappling with what it should stand for in the post-Brexit era, Arm shows that global, industrial leadership does not need billion-pound factories, nor, actually, in a sector attracting generous state support worldwide, billions of pounds.

For a country that has got by for decades with no semiconductor strategy to speak of, mixing just enough cash, decades of experience, and the exuberance of youth is the recipe to begin with.

Arm was a magnet for talent. At first, Acorn hired from Cambridge University, one of the academic institutions at the heart of the UK's knowledge economy. Then, the success of the BBC Micro became a highly effective recruitment tool. Soon after, word got out.

I suspect that trying to create another Arm is as much folly as trying to create the next Google, which has often been the cry from those wishing domestic start-ups could complete the journey to giant, world-beating status. But ministers might lay the groundwork with training and skills, and easing the path to apply brilliant science to all. They could encourage a culture that breeds the boldness to take an American idea, commercialise it, and sell it back to them and the world. And, if domestic ownership is important, they could foster more understanding investors offering patient capital when small firms need space and time to scale up, find new customers and develop new products. More broadly, any successful start-up needs to be international from the get-go. That means well-connected airports, plus a stable tax and regulatory regime and the ability to hire from anywhere.

For now, Arm appears indispensable, powering the inexorable growth in computing and offering lessons about the importance of scale, longevity, continuous innovation and perhaps even diplomacy. But new technologies offer as much threat as opportunity. This industry is littered with breakthroughs that never find a market and standards, systems and devices that are undercut and abandoned when something better comes along.

All the next upstart needs is a simple idea: one that is sold with confidence, capable of riding its luck, adapting to survive, developing predictably and becoming a hub for widespread collaboration. That's how Arm succeeded.

# REFERENCES

## 1. EVERYTHING, EVERYWHERE: HOW THE MICROCHIP TOOK OVER

1   https://www.wsj.com/articles/pelosi-vows-ironclad-defense-of-taiwans-democracy-as-china-plans-live-fire-drills-11659511188

2   https://www.semiconductors.org/study-identifies-benefits-and-vulnerabilities-of-global-semiconductor-supply-chain-recommends-government-actions-to-strengthen-it/

3   https://edition.cnn.com/videos/tv/2022/07/31/exp-gps-0731-mark-liu-taiwan-semiconductors.cnn

4   https://www.futurehorizons.com/assets/fh_research_bulletin_2021-04_the_china_c.pdf

5   https://www.youtube.com/watch?v=FKO5AXIB_Ac

6   https://www.wsj.com/articles/global-chip-shortage-is-far-from-over-as-wait-times-get-longer-11635413402

7   https://news.sky.com/story/ps5-becomes-fastest-sony-console-to-achieve-sales-of-10-million-12366564

8   https://www.reuters.com/business/autos-transportation/boosted-by-premium-car-demand-volkswagen-raises-margin-target-2021-05-06/

9   https://www.alixpartners.com/media-center/press-releases/
    press-release-shortages-related-to-semiconductors-to-cost-
    the-auto-industry-210-billion-in-revenues-this-year-says-
    new-alixpartners-forecast/

10  https://www2.deloitte.com/content/dam/Deloitte/tw/
    Documents/technology-media-telecommunications/tw-
    semiconductor-report-EN.pdf

11  https://www.philips.com/a-w/about/news/archive/standard/
    news/articles/2022/20220608-chips-for-lives-global-chip-
    shortages-put-production-of-life-saving-medical-devices-
    and-systems-at-risk.html

12  https://www.nscai.gov/wp-content/uploads/2021/03/Full-
    Report-Digital-1.pdf

13  https://www2.deloitte.com/content/dam/Deloitte/us/
    Documents/technology-media-telecommunications/
    us-tmt-2022-semiconductor-outlook.pdf

14  https://www.bcg.com/publications/2020/incen-
    tives-and-competitiveness-in-semiconductor-manufacturing

15  https://www.whitehouse.gov/briefing-room/speeches-
    remarks/2021/02/24/remarks-by-president-biden-at-signing-
    of-an-executive-order-on-supply-chains/

16  https://www.nytimes.com/1989/06/14/world/reagan-gets-a-
    red-carpet-from-british.html

17  https://www.semiconductors.org/global-semiconductor-
    sales-units-shipped-reach-all-time-highs-in-2021-as-
    industry-ramps-up-production-amid-shortage/

18  https://docs.cdn.yougov.com/w2zmwpzsq0/econTabReport.pdf

19  https://arxiv.org/pdf/2011.02839.pdf

20  https://www.youtube.com/watch?v=3jU_YhZ1NQA-
    &t=7207s

21  FutureHorizons, Research Brief 09/2021, 'Back to a Vertical
    Business Model'.

## 2. SOME HISTORY: THE ODD COUPLE STARTS OUT

1 https://worldradiohistory.com/Archive-Electronics/60s/61/Electronics-1961-03-10.pdf
2 Displayed over the Coliseum's four floors was the latest in radio electronics, radar, complex air traffic control and space communications.
3 T.R. Reid, *The Chip,* Simon & Schuster, 1985, p. 18.
4 https://everything2.com/title/The%2520Tyranny%2520of%-2520Numbers
5 https://davidlaws.medium.com/the-computer-chip-is-sixty-36cff1d837a1
6 https://www.nobelprize.org/prizes/physics/2000/kilby/biographical/
7 https://www.lindahall.org/about/news/scientist-of-the-day/geoffrey-dummer
8 T.R. Reid, *The Chip*, p. 87.
9 https://digitalassets.lib.berkeley.edu/roho/ucb/text/rock_arthur.pdf
10 Leslie Berlin, *The Man Behind the Microchip*, OUP, 2005, p. 61.
11 Michael Malone, *The Intel Trinity*, HarperCollins, 2014, p. 15.
12 https://digitalassets.lib.berkeley.edu/roho/ucb/text/rock_arthur.pdf
13 T.R. Reid, *The Chip*, p. 147.
14 https://newsroom.intel.com/wp-content/uploads/sites/11/2018/05/moores-law-electronics.pdf
15 https://www.youtube.com/watch?v=EzyJxAP6AQo
16 https://edtechmagazine.com/k12/article/2012/11/calculating-firsts-visual-history-calculators
17 https://www.intel.com/content/www/us/en/history/virtual-vault/articles/intels-founding.html

18  https://digitalassets.lib.berkeley.edu/roho/ucb/text/rock_arthur.pdf

19  https://www.intel.sg/content/www/xa/en/history/museum-story-of-intel-4004.html

## 3: FROM A TINY ACORN, DESIGNS ON THE FUTURE

1  'Their bits are worse than their bytes', United Press International, 24 December 1984.

2  https://www.youtube.com/watch?v=jtMWEiCdsfc

3  http://34.242.82.140/media/BBC-Microelectronic-government-submission.pdf

4  http://nottspolitics.org/wp-content/uploads/2013/06/Labours-Plan-for-science.pdf

5  https://clp.bbcrewind.co.uk/media/BBC-Microelectronic-government-submission.pdf

6  https://www.bbc.co.uk/news/technology-15969065

7  https://www.youtube.com/watch?v=KrTmvqwpZF8

8  Brian Merchant, *The One Device*, Little, Brown, 2017, p. 155.

9  https://archive.computerhistory.org/resources/access/text/2012/06/102746190-05-01-acc.pdf

10  https://media.nesta.org.uk/documents/the_legacy_of_bbc_micro.pdf

11  https://www.nytimes.com/1962/11/03/archives/pocket-computer-may-replace-shopping-list-inventor-says-device.html

12  https://archive.computerhistory.org/resources/access/text/2014/08/102739939-05-01-acc.pdf

13  https://archive.computerhistory.org/resources/access/text/2014/08/102739939-05-01-acc.pdf

14  https://web.archive.org/web/20120721114927/http://www.variantpress.com/view.php?content=ch001

15  https://web.archive.org/web/20120721114927/http://www.
    variantpress.com/view.php?content=ch001

16  Steve Wozniak, *iWoz*, Headline, 2007

17  Walter Isaacson, *Steve Jobs*, Simon & Schuster, 2011, p. 58.

18  https://en.wikipedia.org/wiki/Apple_II#/media/File:Apple_II_
    advertisement_Dec_1977_page_2.jpg

19  Berlin, *The Man Behind the Microchip*, p. 251.

20  Isaacson, *Steve Jobs*, p. 84.

21  https://archive.computerhistory.org/resources/access/tex-
    t/2014/08/102746675-05-01-acc.pdf

22  https://media.nesta.org.uk/documents/the_legacy_of_bbc_
    micro.pdf

23  http://www.naec.org.uk/organisations/bbc-computer-literacy-
    project/towards-computer-literacy-the-bbc-computer-literacy-
    project-1979-1983

24  https://www.margaretthatcher.org/document/104609

25  https://www.theregister.com/2011/11/30/bbc_micro_
    model_b_30th_anniversary/?page=5

26  https://www.youtube.com/watch?v=T2VfgtTt5So

27  1985, January 13. Acorn's star fades. *The Sunday Times*. Bird, J.
    1985, February 10. Apple puts the Bite on Acorn. *The Sunday
    Times*.

28  https://www.express.co.uk/expressyourself/113527/Battle-of-
    the-Boffins

29  Horton, J., Letter to John Biggs, 1984

30  https://archive.computerhistory.org/resources/access/
    text/2016/07/102737949-05-01-acc.pdf

31  https://inst.eecs.berkeley.edu/~n252/paper/RISC-patterson.pdf

32  https://archive.computerhistory.org/resources/access/
    text/2016/07/102737949-05-01-acc.pdf

33  https://www.commodore.ca/commodore-history/the-rise-of-
    mos-technology-the-6502/

34  https://archive.computerhistory.org/resources/access/text/2012/06/102746190-05-01-acc.pdf

35  https://archive.org/details/AcornUser039-Oct85/page/n8/mode/1up

36  Isaacson, *Steve Jobs*, p. 144.

37  https://www.mprove.de/visionreality/media/Kay72a.pdf

38  John Sculley, *Odyssey*, HarperCollins, 1987, pp. 403–4.

39  https://archive.computerhistory.org/resources/access/text/2014/08/102746675-05-01-acc.pdf

40  https://www.nomodes.com/LinzmayerBook.html

41  https://www.nomodes.com/LinzmayerBook.html

42  Sculley, *Odyssey*, p. 342.

## 4. THIRTEEN MEN EMBEDDED IN A BARN

1   https://www.sbsummertheatre.com/history

2   https://docplayer.net/103199502-All-contributions-for-next-month-s-issue-are-required-by-19th-of-each-month-please-send-to-the-edito-r.html

3   https://www.youtube.com/watch?v=ljbdhICqETE

4   'INMOS becomes member of SGS-THOMSON Group', *Business Wire*, 6 April 1989.

5   https://www.youtube.com/watch?v=ljbdhICqETE

6   https://archive.computerhistory.org/resources/access/text/2020/02/102706882-05-01-acc.pdf

7   https://archive.computerhistory.org/resources/access/text/2020/02/102706882-05-01-acc.pdf

8   http://www.nomodes.com/LinzmayerBook.html

9   https://archive.computerhistory.org/resources/access/text/2013/04/102746578-05-01-acc.pdf

10  https://archive.computerhistory.org/resources/access/text/2020/02/102706882-05-01-acc.pdf

11 'Arming the World', *Electronic Business*, 1999.
12 Company brochure supplied by John Biggs.
13 https://www.managementtoday.co.uk/andrew-davidson-interview-robin-saxby-chairman-arm-soaring-microchip-design-company-says-just-following-hobby-success-part-gadget-obsessed-mr-fixit/article/412279

## 5. NOKIA'S MAD PHONE SETS THE STANDARD

1 https://www.nokia.com/blog/thirty-years-on-from-the-call-that-transformed-how-we-communicate/#:~:text=The%20first%20official%20GSM%20call,nights%20to%20make%20it%20happen
2 https://web.archive.org/web/20070213045903/http://telemuseum.no/mambo/content/view/29/1/
3 https://money.cnn.com/magazines/fortune/fortune_archive/2000/05/01/278948/index.htm
4 https://money.cnn.com/1999/02/08/europe/nokia/#:~:-text=Nokia%20overtakes%20Motorola%20%2D%20Feb.,8%2C%201999&text=LONDON%20(CNNfn)%20%2D%20Worldwide%20sales,according%20to%20a%20report%20Monday

## 6. AS ARM CASHES IN, APPLE CASHES OUT

1 https://www.youtube.com/watch?v=QhhFQ-3w5tE
2 https://www.annualreports.com/HostedData/Annual-ReportArchive/a/NASDAQ_AAPL_1996.pdf
3 Isaacson, *Steve Jobs*, p. 276.
4 https://www.youtube.com/watch?v=IOs6hnTI4lw
5 https://www.youtube.com/watch?v=IOs6hnTI4lw

6   https://www.youtube.com/watch?v=qccG0bEB-jYM&list=WL&index=12

7   Isaacson, *Steve Jobs*, p. 283.

8   https://www.zdnet.com/article/newton-inc-apple-spins-off-the-messagepad/

9   https://www.cultofmac.com/469567/today-in-apple-history-apple-bids-farewell-to-the-newton/

10  https://www.4corn.co.uk/articles/websites/www95/acorn/library/pr/1995/07_Jul/NewMD.html

11  https://www.wired.com/2009/12/fail-oracle/

12  https://techmonitor.ai/technology/arm_wins_billion_dollar_valuation_in_ipo

13  Company brochure supplied by John Biggs.

14  https://www.youtube.com/watch?v=i5f8bqYYwps&list=WL

15  https://thenextweb.com/news/ex-apple-ceo-john-sculley-tells-story-arm-newton-start-apple-mobile-giant

16  https://www.nomodes.com/LinzmayerBook.html

17  https://archive.computerhistory.org/resources/access/text/2020/02/102706882-05-01-acc.pdf

18  Acorn Group PLC, 17 March 1998, stock-market statement.

19  https://www.investegate.co.uk/arm-holdings-plc/rns/4th-quarter---final-results-to-31-december-1999/200001310701286413E/

20  https://www.sec.gov/Archives/edgar/data/1057997/000095010303001446/jun2303_20f.htm

21  https://www.theguardian.com/technology/2000/jul/18/efinance.business1

## 7. GOING GLOBAL: HOW ASIA MADE THE MODERN MICROCHIP INDUSTRY

1   https://www.history.com/this-day-in-history/earth-quake-kills-thousands-in-taiwan#:~:text=An%20

earthquake%20in%20Taiwan%20on,tremor%20that%20
killed%203%2C200%20people.

2   https://www.nytimes.com/2000/02/01/business/the-silicon-
    godfather-the-man-behind-taiwan-s-rise-in-the-chip-
    industry.html

3   https://pr.tsmc.com/english/news/2191

4   https://pr.tsmc.com/english/news/2213

5   https://archive.computerhistory.org/resources/access/
    text/2017/03/102740002-05-01-acc.pdf

6   https://archive.computerhistory.org/resources/access/
    text/2017/03/102740002-05-01-acc.pdf

7   https://www.nber.org/system/files/working_papers/w0118/
    w0118.pdf, p. 60.

8   https://www.semi.org/en/Oral-History-Interview-Ed-Pausa

9   https://archive.computerhistory.org/resources/access/
    text/2012/04/102658284-05-01-acc.pdf

10  Berlin, *The Man Behind the Microchip*, p. 132.

11  https://www.nber.org/system/files/working_papers/w0118/
    w0118.pdf

12  https://www.alamy.com/stock-photo-1960s-advertisement-
    advertising-portable-transistor-radios-by-sony-147924076.
    html

13  https://documents1.worldbank.org/curated/
    en/975081468244550798/pdf/multi-page.pdf

14  Geoffrey Cain, *Samsung Rising*, Virgin Books, 2020, p. 53.

15  https://documents1.worldbank.org/curated/
    en/975081468244550798/pdf/multi-page.pdf

16  https://www.hpmemoryproject.org/timeline/art_fong/chuck_
    house_thoughts.htm

17  https://www.nytimes.com/1982/02/28/business/japan-s-big-
    lead-in-memory-chips.html

18 Andy Grove, *Only the Paranoid Survive*, HarperCollins Business, 1996, p. 89.

19 https://archive.computerhistory.org/resources/access/text/2017/03/102740002-05-01-acc.pdf

20 Malone, *The Intel Trinity*, p. 409.

21 https://link.springer.com/chapter/10.1007/4-431-28916-X_3

22 https://www.semi.org/en/Oral-History-Interview-Morris-Chang

23 https://spectrum.ieee.org/morris-chang-foundry-father

24 https://taiwantoday.tw/news.php?unit=6,23,45,6,6&post=8429

25 https://citeseerx.ist.psu.edu/viewdoc/download?-doi=10.1.1.548.6098&rep=rep1&type=pdf

26 https://mediakron.bc.edu/edges/case-studies-in-the-taiwan-miracle/hsinchu-science-park-a-case-study-in-taiwans-shift-to-tech/from-industry-to-innovation-hsinchu-science-park-tsmc-and-the-development-of-taiwans-tech-sector/opening-hsinchu-science-park

27 https://www.semi.org/en/Oral-History-Interview-Morris-Chang

28 https://www.semi.org/en/Oral-History-Interview-Morris-Chang

29 https://www.youtube.com/watch?v=wEh3ZgbvBrE&t=8s

30 https://www.brookings.edu/wp-content/uploads/2022/04/Vying-for-Talent-Morris-Chang-20220414.pdf

31 https://www.youtube.com/watch?v=wEh3ZgbvBrE&t=8s

32 Saxenian and Hsu, 'The Silicon Valley-Hsinchu Connection: Technical Communities and Industrial Upgrading'.

## 8. INTEL INSIDE: A PC POWERHOUSE TRIES TO MOBILISE

1 Shareholders' Meeting – Final, FD (Fair Disclosure) Wire, May 18, 2005 Wednesday transcript.

2   Shareholders' Meeting – Final, FD (Fair Disclosure) Wire, May 18, 2005 Wednesday transcript.

3   Shareholders' Meeting – Final, FD (Fair Disclosure) Wire, May 18, 2005 Wednesday transcript.

4   https://www.intel.com/content/www/us/en/history/history-2004-annual-report.htm

5   file:///C:/Users/james/Downloads/history-2004-annual-report.pdf

6   Malone, *The Intel Trinity*, p. 364.

7   https://www.latimes.com/archives/la-xpm-1991-08-30-fi-1556-story.html

8   Jeff Ferry, 'The best chip shop in the world', *Director*, March 1994.

9   https://www.intel.com/pressroom/kits/events/idffall_2005/20050823Otellini.pdf (Link since taken down by Intel)

10  https://www.intel.com/pressroom/kits/events/idffall_2005/20050823Otellini.pdf

11  https://www.forbes.com/global/2007/0604/062.html?sh=6a85ceffd4f2

12  Malone, *The Intel Trinity*, p. 211.

13  https://arstechnica.com/gadgets/2017/06/ibm-pc-history-part-1/

14  https://www.ibm.com/ibm/history/exhibits/pc25/pc25_intro.html

15  https://spectrum.ieee.org/how-the-ibm-pc-won-then-lost-the-personal-computer-market

16  https://www.ibm.com/ibm/history/ibm100/us/en/icons/personalcomputer/

17  Malone, *The Intel Trinity*, p. 393.

18  https://www.informationweek.com/it-life/ibm-s-elephant-that-couldn-t-tap-dance-with-the-pc

19 'Microsoft Trial – Gates' Spat With Intel Is Revealed By E-Mail', *Seattle Times*, 23 June 1999.

20 https://rarehistoricalphotos.com/windows-95-launch-day-1995/

21 https://www.intel.com/content/www/us/en/history/virtual-vault/articles/end-user-marketing-intel-inside.html

22 'In the Spotlight; The Intel Hustle', *Los Angeles Times*, 7 September 1997.

23 https://books.google.co.uk/books?id=MTd-CDwAAQBAJ&pg=PA1411&lpg=PA1411&dq=intel+p-c+market+share+56+1989&source=bl&ots=Yg-CRw-J9gn&sig=ACfU3U32KeS_uKlRZrnpNzFkgeyqJMK8Y-w&hl=en&sa=X&ved=2ahUKEwiVxoaypb74AhVJXsAKH-QgxBYcQ6AF6BAgyEAM#v=onepage&q=intel%20pc%20market%20share%2056%201989&f=false

24 https://queue.acm.org/detail.cfm?id=957732

25 'Arm Ltd Partners with Digital – Acorn joint venture company continues run of success', M2 PRESSWIRE, 21 February 1995.

26 https://archive.computerhistory.org/resources/access/text/2014/01/102746627-05-01-acc.pdf

## 9. SPACE INVADERS AND THE RACE FOR THE IPHONE

1 https://www.ingenia.org.uk/ingenia/issue-69/profile

2 http://media.corporate-ir.net/media_files/irol/19/197211/626-1_Arm_AR_040311.pdf

3 https://www.quora.com/What-was-it-like-working-on-the-original-iPhone-project-codenamed-Project-Purple

4 Merchant, *The One Device*, p. 224.

5 https://www.youtube.com/watch?v=cp49Tmmtmf8

6 Merchant, *The One Device*, p. 150.

7 Cain, *Samsung Rising*, p. 60.

8  https://techcrunch.com/2008/02/27/over-a-billion-mobile-phones-sold-in-2007/

9  Cain, *Samsung Rising*, p. 152.

10  https://www.youtube.com/watch?v=MnrJzXM7a6o

11  https://www.globenewswire.com/en/news-release/2004/10/19/317281/2693/en/Arm-Introduces-The-Cortex-M3-Processor-To-Deliver-High-Performance-In-Low-Cost-Applications.html

12  'Chip Off Silicon Valley's Block', *The Business*, 29 August 2004.

13  https://www.telegraph.co.uk/finance/2893316/Arm-shares-fall-18pc-after-US-acquisition.html

14  https://www.intel.com/pressroom/archive/releases/2006/20060627corp.htm

15  https://www.theguardian.com/technology/2009/jun/11/intel-culv-sean-maloney

16  Grove, *Only the Paranoid Survive*, p. 105.

17  https://appleinsider.com/articles/07/12/21/exclusive_apple_to_adopt_intels_ultra_mobile_pc_platform.html

18  Isaacson, *Steve Jobs*, p. 454.

19  Isaacson, *Steve Jobs*, pp. 454–5.

20  Q4 2007 Arm Holdings plc Earnings Presentation – Final, FD (Fair Disclosure) Wire, 5 February 2008 (nexis.com)

21  https://fortune.com/2009/07/16/the-chip-company-that-dares-to-battle-intel/

22  https://www.annualreports.com/HostedData/Annual-ReportArchive/i/NASDAQ_INTC_2012.pdf

23  https://www.annualreports.com/HostedData/Annual-ReportArchive/a/NASDAQ_AAPL_2012.pdf

24  https://www.theatlantic.com/technology/archive/2013/05/paul-otellinis-intel-can-the-company-that-built-the-future-survive-it/275825/

25  https://www.arm.com/company/news/2013/02/arm-hold-ings-reports-results-for-fourth-quarter-and-full-year-2012

26  https://www.theatlantic.com/technology/archive/2013/05/paul-otellinis-intel-can-the-company-that-built-the-future-survive-it/275825/

27  https://newsroom.intel.com/editorials/brian-krzanich-our-strategy-and-the-future-of-intel/#gs.lecequ

28  https://www.apple.com/uk/newsroom/2020/06/apple-announces-mac-transition-to-apple-silicon/

## 10. A 300-YEAR VISION, A 64-DAY TAKEOVER

1   Duncan Clark, *Alibaba: The House that Jack Ma Built*, Ecco, 2016.

2   https://www.reuters.com/article/alibaba-ipo-board-idINL-4N0QK3Q120140827

3   Atsuo Inoue, *Aiming High: Masayoshi Son, SoftBank, and Disrupting Silicon Valley*, Hodder & Stoughton, 2021.

4   https://www.independent.co.uk/news/people/profiles/simon-segars-interview-looking-forward-future-and-internet-things-9789959.html

5   https://www.independent.co.uk/news/people/profiles/simon-segars-interview-looking-forward-future-and-internet-things-9789959.html

6   https://www.investegate.co.uk/arm-holdings-plc--arm-/rns/analyst-and-investor-day-2015/201509150700080227Z/

7   https://www.mckinsey.com/business-functions/mckinsey-digital/our-insights/the-internet-of-things-the-value-of-digitizing-the-physical-world

8   https://asia.nikkei.com/Business/Companies/Masayoshi-Son-talks-about-how-Steve-Jobs-inspired-SoftBank-s-Arm-deal

9   Inoue, *Aiming High*, p. 270.

10  https://www.businesswire.com/news/home/20160621005758/en/SoftBank-to-Sell-Supercell-Stake-at-USD-10.2-Billion-Valuation

11  https://group.softbank/en/news/press/20150511_4

12  'SoftBank CEO Son plans to work at least 5 more yrs as Arora quits', *Japan Economic Newswire*, 22 June 2016.

13  https://www.livemint.com/Companies/uzZ0D4e4DyjvqI-UqETMbYP/The-trigger-for-Nikesh-Aroras-SoftBank-resignation.html

14  https://www.deepchip.com/items/0562-04.html

15  https://www.theresa2016.co.uk/we_can_make_britain_a_country_that_works_for_everyone

16  https://www.youtube.com/watch?v=ZzhYOPIelb4

17  https://www.enterprise.cam.ac.uk/a-call-to-arms/

18  https://www.youtube.com/watch?v=d1S7Zk3eHdo

19  https://www.techinasia.com/masayoshi-son-softbank-40-year-dream-arm-acquisition

20  https://asia.nikkei.com/Business/Companies/Masayoshi-Son-talks-about-how-Steve-Jobs-inspired-SoftBank-s-Arm-deal

21  https://asia.nikkei.com/Business/Companies/Masayoshi-Son-talks-about-how-Steve-Jobs-inspired-SoftBank-s-Arm-deal

22  Personal view; British fund managers are too risk averse to back visionaries such as Elon Musk, writes James Anderson. *The Daily Telegraph* (London) 15 April 2017

23  https://www.youtube.com/watch?v=ZzhYOPIelb4

## 11. GOING GLOBAL: CHINA PLAYS ITS TRUMP CARD

1   https://www.youtube.com/watch?v=1Z_ZcMdYRrA

2   https://www.youtube.com/watch?v=1Z_ZcMdYRrA

3   https://investors.broadcom.com/news-releases/news-release-details/broadcom-proposes-acquire-qualcomm-7000-share-cash-and-stock-0

4   https://www.sec.gov/Archives/edgar/data/804328/000110465918015036/a18-7296_7ex99d1.htm

5   https://www.sec.gov/Archives/edgar/data/804328/000110465918015036/a18-7296_7ex99d1.htm

6   https://www.everycrsreport.com/reports/RL33388.html

7   https://www.nytimes.com/1987/03/17/business/japanese-purchase-of-chip-maker-canceled-after-objections-in-us.html

8   https://www.bbc.com/news/business-43380893

9   https://trumpwhitehouse.archives.gov/wp-content/uploads/2017/12/NSS-Final-12-18-2017-0905.pdf

10  https://obamawhitehouse.archives.gov/blog/2017/01/09/ensuring-us-leadership-and-innovation-semiconductors

11  https://www.reuters.com/article/nxp-semicondtrs-ma-qualcomm-mollenkopf-idUSL1N1UN01L

12  https://www.reuters.com/article/nxp-semicondtrs-ma-qualcomm-idUSFWN1Y704B

13  https://www.nanya.com/en/About/27/Corporate%20Milestone

14  https://www.youtube.com/watch?v=mKzMYgOE6sw

15  https://www.youtube.com/watch?v=mKzMYgOE6sw

16  https://www.mckinsey.com/industries/semiconductors/our-insights/semiconductors-in-china-brave-new-world-or-same-old-story

17  From target-setting advisory document produced by the 'China Manufacturing Commission', seen by the author.

18  https://www.mckinsey.com/featured-insights/asia-pacific/a-new-world-under-construction-china-and-semiconductors

19 https://www.patentlyapple.com/patently-apple/2017/10/
apples-coo-jeff-williams-recounts-how-business-with-tsmc-
began-with-a-dinner-at-the-founders-home.html

20 https://phys.org/news/2009-11-china-chip-maker-mln-tsmc.
html

21 https://www.pwc.com/gx/en/technology/pdf/china-semicon-
2015-report-1-5.pdf

22 https://asia.nikkei.com/Business/China-tech/Taiwan-loses-
3-000-chip-engineers-to-Made-in-China-2025

23 https://www.ft.com/content/8e6271aa-a1d1-4ddc-8b94-
8480c9cb3ce0

24 https://www.independent.co.uk/news/business/analysis-and-
features/huawei-founder-brushes-off-accusations-that-it-
acts-as-an-arm-of-the-chinese-state-9319244.html

25 https://www.fiercewireless.com/wireless/huawei-equipment-
currently-deployed-by-25-u-s-rural-wireless-carriers-rwa-says

26 https://www.gsmarena.com/huaweis_2018_reve-
nue_surpasses_100_million_for_the_first_time__-
news-36279.php

27 https://chinacopyrightandmedia.wordpress.
com/2016/04/19/speech-at-the-work-conference-for-cyber-
security-and-informatization/

28 https://www.politico.com/story/2016/06/full-transcript-
trump-job-plan-speech-224891

29 https://www.politifact.com/article/2016/jul/01/donald-
trump-cites-ronald-reagan-protectionist-her/

30 https://www.handelsblatt.com/technik/it-internet/interview-
huawei-founder-ren-zhengfei-5g-is-like-a-nuclear-bomb-for-
the-us/24240894.html

31 https://www.bbc.co.uk/news/technology-48363772

32 https://www.reuters.com/article/us-asml-holding-usa-chi-
na-insight-idUSKBN1Z50HN

33 https://2017-2021.commerce.gov/news/press-releases/2020/12/commerce-adds-chinas-smic-entity-list-restricting-access-key-enabling.html

34 https://www.semi.org/en/news-media-press/semi-press-releases/semi-export-control

35 https://www.reuters.com/article/us-southkorea-japan-laborers-analysis-idUSKCN1U31GS

## 12. BIG DATA AND WALKING WITH GIANTS

1 https://www.theguardian.com/environment/2020/jan/06/why-irish-data-centre-boom-complicating-climate-efforts

2 https://www.cso.ie/en/releasesandpublications/ep/p-dcmec/datacentresmeteredelectricityconsumption2020/keyfindings/

3 https://www.datacenterdynamics.com/en/news/apple-declines-to-commit-to-galway-data-center-irish-govt-promises-to-do-anything/

4 https://www.eirgridgroup.com/newsroom/all-island-gcs-2019/

5 https://www.telegraph.co.uk/luxury/technology/tech-insiders-really-think-andy-jassy-soon-to-be-ceo-amazon/

6 https://ir.aboutamazon.com/news-release/news-release-details/2020/Amazoncom-Announces-Fourth-Quarter-Sales-up-21-to-874-Billion/default.aspx

7 https://www.gartner.com/en/newsroom/press-releases/2021-06-28-gartner-says-worldwide-iaas-public-cloud-services-market-grew-40-7-percent-in-2020

8 https://www.youtube.com/watch?v=7-31KgImGgU

9 https://www.youtube.com/watch?v=7-31KgImGgU

10 Arm Holdings plc Annual Report 2015: Strategic Report

11  https://aws.amazon.com/blogs/compute/15-years-of-sili-con-innovation-with-amazon-ec2/

12  https://www.forbes.com/consent/?toURL=https://www.forbes.com/2008/04/23/apple-buys-pasemi-tech-ebiz-cz_eb_0422apple.html

13  Event Brief of Q1 2008 Arm Holdings plc Earnings Conference Call – Final FD (Fair Disclosure) Wire 29 April 2008.

14  Sculley, *Odyssey*, p. 163.

15  https://www.cultofmac.com/484394/apple-intel-over-powerpc/

16  https://www.bloomberg.com/features/2016-johny-srouji-apple-chief-chipmaker/

17  https://www.apple.com/uk/newsroom/2010/01/27Apple-Launches-iPad/#:~:text=SAN%20FRANCISCO%20%E2%80%94%20January%2027th%2C%202010,e%2D-books%20and%20much%20more

18  https://www.youtube.com/watch?v=zZtWlSDvb_k

19  Isaacson, *Steve Jobs*, p. 472.

20  https://www.mckinsey.com/industries/semiconductors/our-insights/whats-next-for-semiconductor-profits-and-value-creation

21  https://www.bloomberg.com/news/articles/2020-12-10/apple-starts-work-on-its-own-cellular-modem-chip-chief-says#x-j4y7vzkg

22  https://blog.google/products/pixel/introducing-google-tensor/

23  https://techcrunch.com/2019/04/22/tesla-vaunts-creation-of-the-best-chip-in-the-world-for-self-driving/?guccounter=1&guce_referrer=aHR0cHM6Ly93d-3cuZ29vZ2xlLmNvbS88&guce_referrer_sig=AQAAADm-6BrFV203VRHRLQndFf8gJBYJK7pz_ovVs--TO52FsyxbN-G3rCKZ4rZMLRVd7raYmcX

24  https://www.counterpointresearch.com/semiconductor-reve-nue-ranking-2021/

## 13. NVIDIA'S PARALLEL UNIVERSE

1   https://group.softbank/en/news/webcast/20190919_01_en
2   https://group.softbank/en/news/press/20161014
3   https://www.wework.com/newsroom/wecompany
4   https://www.reuters.com/article/us-softbank-group-results-idUSKBN1XG0Q9
5   https://group.softbank/system/files/pdf/ir/financials/annual_reports/annual-report_fy2020_01_en.pdf
6   https://group.softbank/system/files/pdf/ir/financials/annual_reports/annual-report_fy2020_01_en.pdf
7   https://group.softbank/en/news/webcast/20190919_02_en
8   https://group.softbank/en/news/webcast/20190919_02_en
9   https://www.gartner.com/en/newsroom/press-releases/2014-11-11-gartner-says-nearly-5-billion-connected-things-will-be-in-use-in-2015
10  https://www.gartner.com/en/newsroom/press-releases/2016-01-14-gartner-says-by-2020-more-than-half-of-major-new-business-processes-and-systems-will-incorporate-some-element-of-the-internet-of-things
11  https://www.gartner.com/en/newsroom/press-releases/2017-02-07-gartner-says-8-billion-connected-things-will-be-in-use-in-2017-up-31-percent-from-2016
12  https://group.softbank/system/files/pdf/ir/financials/annual_reports/annual-report_fy2020_01_en.pdf
13  https://www.electronicsweekly.com/blogs/mannerisms/dilemmas/arm-ipo-2023-2019-10/
14  https://group.softbank/system/files/pdf/ir/financials/annual_reports/annual-report_fy2017_01_en.pdf
15  'Accomplished team of graphics and multimedia experts', *Business Wire*, 25 July 1994.
16  https://pressreleases.responsesource.com/news/3992/nvidia-launches-the-world-s-first-graphics-processing-unit-geforce-256/

17 http://www.machinelearning.org/archive/icml2009/papers/218.pdf

18 https://www.telegraph.co.uk/technology/2020/09/19/nvidia-boss-vows-protect-arm-generation-company/

19 https://group.softbank/en/news/press/20200914_0

20 https://blogs.nvidia.com/blog/2020/09/13/jensen-employee-letter-arm/

21 https://asia.nikkei.com/Business/SoftBank2/SoftBank-s-Son-entrusts-Arm-to-Nvidia-s-leather-jacket-clad-chief

22 Inoue, *Aiming High*, p. 271.

23 https://www.telegraph.co.uk/technology/2020/09/19/nvidia-boss-vows-protect-arm-generation-company/

24 https://group.softbank/en/news/press/20180605

25 https://asia.nikkei.com/Business/Companies/Arm-s-China-joint-venture-ensures-access-to-vital-technology

26 https://asia.nikkei.com/Business/China-tech/How-SoftBank-s-sale-of-Arm-China-sowed-the-seeds-of-discord

27 https://www.bloomberg.com/news/articles/2021-02-12/google-microsoft-qualcomm-protest-nvidia-s-arm-acquisition

28 https://www.savearm.co.uk/

29 https://www.theresa2016.co.uk/we_can_make_britain_a_country_that_works_for_everyone

30 https://www.bbc.co.uk/news/business-52275201

31 https://www.arm.com/company/news/2021/03/arms-answer-to-the-future-of-ai-armv9-architecture#:~:text=Cambridge%2C%20UK%2C%20March%2030%2C,and%20artificial%20intelligence%20(AI).

32 https://nvidianews.nvidia.com/news/nvidia-launches-uks-most-powerful-supercomputer-for-research-in-ai-and-healthcare#:~:text=Cambridge%2D1%20is%20the%20first,to%20the%20greater%20scientific%20community

33  https://www.ftc.gov/news-events/news/press-releases/2021/12/ftc-sues-block-40-billion-semiconductor-chip-merger

34  https://assets.publishing.service.gov.uk/media/61d8-1a458fa8f505953f4ed7/NVIDIA-Arm_-_CMA_Initial_Submission_-_NCV_for_publication__Revised_23_December_2021_.pdf

35  https://www.theatlantic.com/technology/archive/2013/05/paul-otellinis-intel-can-the-company-that-built-the-future-survive-it/275825/

36  https://assets.publishing.service.gov.uk/media/61d8-1a458fa8f505953f4ed7/NVIDIA-Arm_-_CMA_Initial_Submission_-_NCV_for_publication__Revised_23_December_2021_.pdf

37  https://group.softbank/en/news/press/20220208

38  https://www.arm.com/company/news/2022/02/arm-appoints-rene-haas-as-ceo

## 14. THE RISK FROM RISC-V

1  https://assets.publishing.service.gov.uk/media/61d8-1a458fa8f505953f4ed7/NVIDIA-Arm_-_CMA_Initial_Submission_-_NCV_for_publication__Revised_23_December_2021_.pdf

2  RISC-V International (2020) RISC-V 10th Anniversary: Founders Reflect on RISC-V's Past and Future. Available at: https://www.youtube.com/watch?v=V7fuE1yXUxk

3  https://www2.eecs.berkeley.edu/Pubs/TechRpts/2014/EECS-2014-146.pdf

4  https://assets.publishing.service.gov.uk/media/61d8-1a458fa8f505953f4ed7/NVIDIA-Arm_-_CMA_Initial_Submission_-_NCV_for_publication__Revised_23_December_2021_.pdf

5   https://archive.computerhistory.org/resources/access/
    text/2016/07/102737949-05-01-acc.pdf

6   https://github.com/arm-facts/arm-basics.com/blob/master/
    index.md

7   https://www.reuters.com/article/us-usa-china-semiconduc-
    tors-insight-idUSKBN1XZ16L

8   https://riscv.org/about/history/

9   https://www.reuters.com/article/us-usa-china-semiconduc-
    tors-insight-idUSKBN1XZ16L

10  https://venturebeat.com/2020/09/17/sifive-hires-qualcomm-
    exec-as-ceo-for-risc-v-alternatives-to-nvidia-arm/

11  https://www.sifive.com/blog/sifive-arrives-on-the-pitch-in-
    cambridge

12  https://riscv.org/blog/2022/02/semico-researchs-new-report-
    predicts-there-will-be-25-billion-risc-v-based-ai-socs-
    by-2027/

13  '15m' https://riscv.org/risc-v-10th/

## 15. GOING GLOBAL: SOME FURIOUS SOVEREIGN SPENDING

1   https://www.taiwannews.com.tw/en/news/4360807

2   https://investor.tsmc.com/static/annualReports/2021/
    english/index.html

3   https://www.electronicsweekly.com/news/busi-
    ness/771343-2021-04/

4   https://www.scmp.com/tech/big-tech/article/3130628/tsmc-
    founder-morris-chang-says-chinas-semiconductor-industry-
    still

5   https://www.theregister.com/2022/03/14/taiwan_china_
    tech_worker_raids/

6   https://www.reuters.com/technology/us-considers-crack-
    down-memory-chip-makers-china-2022-08-01/

7  https://press.armywarcollege.edu/cgi/viewcontent.cgi?-article=3089&context=parameters

8  https://www.youtube.com/watch?v=BtYYGcoyWX4

9  https://www.canalys.com/newsroom/Canalys-huawei-samsung-worldwide-smartphone-market-q2-2020?ctid=1556-1195484408fbbb34e0298b96eddb178f

10  https://www.huawei.com/en/news/2022/3/huawei-annual-report-2021

11  https://www.reuters.com/article/huawei-chairman-idCNL1N2TG03F

12  https://www.reuters.com/business/autos-transportation/biden-admin-defends-approving-licenses-auto-chips-huawei-2021-08-27/

13  https://asia.nikkei.com/Spotlight/Huawei-crackdown/Huawei-bets-big-on-chip-packaging-to-counter-U.S.-clampdown

14  https://www.bloomberg.com/news/articles/2022-07-15/huawei-s-secretive-chip-arm-seeks-phds-to-get-past-us-sanctions

15  https://merics.org/en/report/chinas-rise-semiconduc-tors-and-europe-recommendations-policy-makers

16  https://www.semiconductors.org/chinas-share-of-global-chip-sales-now-surpasses-taiwan-closing-in-on-europe-and-japan/

17  https://investor.qualcomm.com/financial-information/sec-filings/content/0001728949-21-000076/0001728949-21-000076.pdf

18  *China chases chip-factory dominance-and global clout, The Wall Street Journal.* Dow Jones & Company. Available at: http://www.wsj.com/articles/china-bets-big-on-basic-chips-in-self-sufficiency-push-11658660402

19  https://www2.deloitte.com/content/dam/Deloitte/us/Documents/technology-media-telecommunications/us-tmt-2022-semiconductor-outlook.pdf

20 https://www.wsj.com/articles/china-bets-big-on-basic-chips-in-self-sufficiency-push-11658660402

21 https://www.techinsights.com/blog/disruptive-technology-7nm-smic-minerva-bitcoin-miner

22 https://www.bis.doc.gov/index.php/documents/about-bis/newsroom/press-releases/3158-2022-10-07-bis-press-release-advanced-computing-and-semiconductor-manufacturing-controls-final/file

23 https://asia.nikkei.com/Spotlight/The-Big-Story/China-s-chip-industry-fights-to-survive-U.S.-tech-crackdown

24 https://www.semiconductors.org/strengthening-the-global-semiconductor-supply-chain-in-an-uncertain-era/

25 https://www.marketwatch.com/story/pat-gelsinger-seeks-to-rescue-intel-in-biggest-return-of-a-prodigal-son-since-steve-jobs-went-back-to-apple-11610570841

26 https://www.intel.com/content/www/us/en/newsroom/news/note-from-pat-gelsinger.html#gs.burq6s

27 https://www.youtube.com/watch?v=gAuh7igXX-s

28 https://www.asml.com/en/investors/annual-report/2021

29 https://www.asml.com/en/news/stories/2022/technological-sovereignty-in-the-chip-industry

30 https://www.seagate.com/files/www-content/our-story/trends/files/idc-seagate-dataage-whitepaper.pdf

31 https://spectrum.ieee.org/gordon-moore-the-man-whose-name-means-progress

32 https://www.whitehouse.gov/briefing-room/speeches-remarks/2022/08/09/remarks-by-president-biden-at-signing-of-h-r-4346-the-chips-and-science-act-of-2022/

33 https://www.finance.senate.gov/imo/media/doc/Louie%20Subcommittee%20Hearing%20on%20International%20Trade%20Customs%20and%20Global%20Competitiveness%20Hearing%205-25-22%20(For%20submission).pdf

34  https://ec.europa.eu/commission/presscorner/detail/en/SPEECH_21_4701

35  https://www.intel.com/content/www/us/en/newsroom/news/eu-news-2022-release.html

36  https://www.theguardian.com/technology/2021/nov/24/samsung-to-build-a-17bn-semiconductor-factory-in-texas-us-chip-shortage

37  https://www.nst.com.my/world/region/2021/05/690183/south-korea-set-worlds-largest-semiconductor-supply-chain

38  https://www.neologicvlsi.com/blog

39  https://www.ft.com/content/b041b2ce-2137-4c14-aa81-9903b29f8978

40  https://www.semiconductors.org/wp-content/uploads/2020/09/Government-Incentives-and-US-Competitiveness-in-Semiconductor-Manufacturing-Sep-2020.pdf

41  https://www.theregister.com/2022/04/20/us_chips_tsmc/

42  https://www.intc.com/news-events/press-releases/detail/1563/intel-reports-second-quarter-2022-financial-results

43  https://www.wsts.org/76/103/The-World-Semiconductor-Trade-Statistics-WSTS-has-released-its-new-semiconductor-market-forecast-generated-in-August-2022

## 16. ALWAYS ON

1  https://www.theguardian.com/business/2022/feb/11/softbank-arm-flotation-legal-fight-china-london-stock-exchange-nasdaq

2  https://www.ft.com/content/43d11498-fd49-4df2-b8ef-aa4a5bbe8852

3  https://www.thisismoney.co.uk/money/markets/article-10950551/SoftBank-boss-dampens-Arm-London-hopes.html

4  https://www.arm.com/company/news/2023/02/arm-announces-q3-fy22-results

5   https://www.datacenterknowledge.com/arm/arm-chips-gaining-data-centers-still-single-digits

6   https://www.tomshardware.com/news/arm-socs-to-grab-30-percent-of-pc-market-by-2026-analyst

7   https://www.ft.com/content/a09c4500-27ae-42d7-8b3f-e6d-13f1b3f3b

8   https://www.ft.com/content/48baeb67-2d3c-41c3-8645-e89ac69a985a

9   https://www.ft.com/content/2fa93008-5a63-4955-b813-c084b617f86b

10  https://group.softbank/en/news/press/20220430

11  https://www.theguardian.com/business/2022/aug/08/softbank-vision-funds-cuts-loss-arm?ref=todayheadlines.live

12  https://www.tomshardware.com/news/arm-sues-qualcomm-and-nuvia-for-breaking-license-agreement

13  https://www.electronicsweekly.com/news/business/intel-interested-consortium-buy-arm-2022-02/

14  https://www.ft.com/content/eab1d19d-ab4c-45b7-88b4-f1f5e115d16e

15  https://www.kedglobal.com/mergers-acquisitions/newsView/ked202209220006

16  https://www.arm.com/company/news/2023/02/arm-announces-q3-fy22-results?Arm%20achieves%20record%20revenue%20and%20shipments%20in%20Q3%20FY%202022

17  http://media.corporate-ir.net/media_files/irol/19/197211/626-1_Arm_AR_040311.pdf

18  Merchant, *The One Device*, p. 161.

19  https://www.arm.com/company/news/2021/03/arms-answer-to-the-future-of-ai-armv9-architecture

# GLOSSARY

*Bit size*

The volume or 'word length' of information that a chip can
handle. A bit is the smallest unit of information in computing, a word blended from 'binary' and 'digit', and usually
represented by a one or a zero.

An 8-bit chip can handle from zero to 255 pieces of data,
which translates from the decimal number system into
11111111 in binary. From there, capacity goes up expo-
nentially. A 16-bit chip handles 0 to 65,535 pieces of data;
a 32-bit chip from 0 to 4,294,967,295; a 64-bit from 0 to
18,446,744,073,709,551,616 pieces. Bits are assembled
in a group of eight to form a byte, a unit used to measure
memory capacity.

*Central processing unit (CPU)*

One type of logic chip, the brains of a computer that processes
the application requests and high-level computer functions
as dictated by software programs.

*Clock cycle*

The heartbeat of a chip and measure of how long it takes one
instruction to be carried out. One megahertz indicates one

million cycles – or instructions – per second. One gigahertz
indicates one billion cycles per second.

*Computer core*

A single processor at the heart of a CPU. Modern CPUs can contain thousands of processors operating on tasks in tandem,
described as a multi-core processor.

*Die*

A small block of semiconducting material on which a microchip is manufactured. Cut from a silicon wafer after bulk
fabrication.

*Digital signal processor (DSP)*

A specialist microprocessor that compresses analogue signals
and is used widely in telecommunications, digital imaging
and speech recognition.

*Embedded system*

A computer combining a processor with memory and some
peripherals that has a dedicated function with a larger
mechanical or electronic system.

*Encapsulation*

A process of setting microchip in resin to protect it from damage.

*Graphics processing unit (GPU)*

A processor specialising in carrying out graphics tasks, but because
it can process many pieces of data simultaneously it has been
applied to machine learning and gaming applications.

*Input/Output (I/O)*

The communication function that connects a processor to the outside world, which could be a human user or another processor.

*Instruction set*

A group of commands for a central processing unit (CPU) in
machine language.

*Integrated circuit (IC)*

A set of electronic circuits on a single piece of semiconducting material, usually silicon. Also known as a chip or microchip.

*Logic chip*

The brains of electronic devices that include CPUs (central processing units), GPUs (graphical processing units) and NPUs (neural processing units, designed for machine learning applications).

*Memory chip*

An information storage unit that works alongside logic chips. Dynamic random access memory (DRAM) was the first memory type, storing small amounts of data often temporarily, so that it could be accessed quickly when a device is turned on. Memory chips freed up the CPU to perform other tasks. Later came slower-moving NAND flash – so called for its similarity to the NOT-AND logic gate and because a section of memory cells could be erased quickly, 'in a flash'. Flash can store greater volumes of data, usually permanently and when a device is switched off, such as a bank of photos carried on a smartphone.

*Microchip*

A set of electronic circuits on a single piece of semiconducting material, usually silicon. Also known as an integrated circuit (IC).

*Microcontroller*

A simpler version of a microprocessor, with integrated memory and peripherals that can take up tasks without recourse to the core. It is used in embedded applications, such as medical devices, remote controls and electrical appliances.

*Microprocessor*

A processor which includes on a single integrated circuit the data processing logic and control required to perform the functions of a computer's central processing unit.

*Peripheral*

An external device that can be attached to a computer to transfer information in or out. Includes mouse, keyboard, webcam, speaker, printer, storage.

*Process node*

A semiconductor manufacturing tier, often measured in nanometres, which refers to the distance between transistors packed onto a microchip. The smaller the number, the faster and more power-efficient the chip.

*Substrate*

The supporting material on which a chip is fabricated or within a chip, or to which a chip is later attached.

*System-on-a-chip*

An integrated circuit featuring all or most of the components of a computer on a single chip.

*Transistor*

A building block of modern electronics, that switches or amplifies electrical signals through logic gates in a circuit.

# ACKNOWLEDGEMENTS

In thanking everyone who spared time to talk to me for this book, or review sections and offer feedback, I am reminded of the saying that you don't know what you don't know.

I was aware my knowledge of microchips was glancing coming into this project. I exit it in awe of the scale of understanding of some of these contributors and hope just a little of that has rubbed off onto me and clung to the pages.

What came across repeatedly was the pride contributors took from their achievements and collective endeavour. And so they should. Arm is a unique company, part of an industry that has truly changed the world.

With thanks to: Nigel and Suzanne Allen, Krste Asanović, John Berylson, John Biggs, Malcolm Bird, Stan Boland, Sir Peter Bonfield, Jonathan Brooks, Tudor Brown, Graham Budd, Stuart Chambers, Richard Conway, Kate Cornish-Bowden, Phil David, Jo De Boeck, Gilles Delfassy, Simon Duke, Doug Dunn, Warren East, Steve Furber, Paul Gavarini, Lord Grimstone of Boscobel, Richard Grisenthwaite, Rene Haas, Alex Harrod, Hermann Hauser, Phil Hughes, Dave Jaggar, David Kershaw, Anastassia Lauterbach, Mark Liu, Craig Livingstone, David Lowdell, Pete Magowan, Chris Malachowsky, Timo Mukari, Malcolm Penn, Tom Pittard, Sudhakar Ram, Ben Rayner, Calista Redmond,

Wally Rhines, Dick Sanquini, Sir Robin Saxby, John Scarisbrick, Simon Segars, Ian Smythe, Chris Thomas, Ian Thornton, David Tupman, Tommi Uhari, Jamie Urquhart, Frits van Hout, Eliza Walsh, Dan Wang, Madeline Wright.

Huw Armstrong and his colleagues at Hodder & Stoughton shared my vision from the off and helped me to make this book what I always hoped it would be. My agent, Toby Mundy at Aevitas Creative Management, deserves great credit for putting us together and offering much support during the process.

I should give special thanks to my in-laws, Anders and Kicki Alvestrand, for the use of 'hönshuset', the perfect writing retreat during one baking summer. And, of course, Viveka and Alice, for their endless love and understanding when another weekend was wiped out – and Oscar, who is always with us.

# BIBLIOGRAPHY

I was spoilt with a choice of reading material for this project, including several sources of excellent oral histories: The Centre for Computing History, Computer History Museum and SEMI.

The Institute of Electrical and Electronics Engineers (IEEE), SemiWiki.com, The Asianometry Newsletter, The Register and *Electronics Weekly* were also enlightening resources.

There is a plethora of books that tackle the broad sweep of industry development, some in much technical detail, plus many useful corporate histories.

Bauer, L.O., Wilder, E.M., *The Microchip Revolution: A Brief History* (2020).

Berlin, L., *The Man Behind the Microchip: Robert Noyce and the Invention of Silicon Valley* (2005).

Cain, G., *Samsung Rising: The Inside Story of the South Korean Giant that Set Out to Beat Apple and Conquer Tech* (2020).

Grove, A.S., *Only the Paranoid Survive: How to Exploit the Crisis Points that Challenge Every Company and Career* (1996).

Inoue, A., *Aiming High: Masayoshi Son, SoftBank, and Disrupting Silicon Valley* (2021).

Isaacson, W., *Steve Jobs* (2011).

Lean, T., *Electronic Dreams: How 1980s Britain Learned to Love the Computer* (2016).

Malone, M.S., *The Intel Trinity: How Robert Noyce, Gordon Moore and Andy Grove Built the World's Most Important Company* (2014).

Mazurek, J., *Making Microchips: Policy, Globalization and Economic Restructuring in the Semiconductor Industry* (1999).

Merchant, B., *The One Device: The Secret History of the iPhone* (2017).

Nenni, D., McLellan, P., *Fabless: The Transformation of the Semiconductor Industry* (2013).

Nenni, D., Dingee, D., *Mobile Unleashed: The Origin and Evolution of Arm Processors in Our Devices* (2015).

Reid, T.R., *The Chip: How Two Americans Invented the Microchip and Launched a Revolution* (2001).

Sculley, J., *Odyssey: Pepsi to Apple . . . A Journey of Adventure, Ideas and the Future* (1987).

# INDEX